Education for Development

An Analysis of Investment Choices

A World Bank Publication

Education
for
Development

An Analysis of Investment Choices

George Psacharopoulos

Maureen Woodhall

Published for The World Bank

Oxford University Press

Oxford University Press

NEW YORK OXFORD LONDON GLASGOW
TORONTO MELBOURNE WELLINGTON HONG KONG
TOKYO KUALA LUMPUR SINGAPORE JAKARTA
DELHI BOMBAY CALCUTTA MADRAS KARACHI
NAIROBI DAR ES SALAAM CAPE TOWN

Library of Congress Cataloging in Publication Data

Psacharopoulos, George.
  Education for development.

  Bibliography: p.
  Includes indexes.
  1. Economic development—Effect of education on.
2. Education—Developing countries—Finance. I. Woodhall,
Maureen. II. Title.
HD75.7.P77 1985   338.9   85-13782
ISBN 0-19-520477-8
ISBN 0-19-520478-6 (pbk.)

Foreword

INVESTMENT IN EDUCATION is a key element of the development process. Its importance is reflected in the growing recognition, since the early 1960s, that investing in both formal and informal education and training provides and enhances the skills, knowledge, attitudes, and motivation necessary for economic and social development. For more than twenty years, the World Bank has been lending for education in developing countries, and experience has been accumulating with respect to the formulation of educational policies and the design of individual projects. This experience not only underlines the importance of educational investment, but also demonstrates the complexity of designing educational policies and investment strategies so they will make the maximum contribution to a country's development effort. The elaboration of appropriate policies necessarily draws on the resources of different scientific disciplines and relies on many analytical tools and a wide variety of information.

The fundamental issues that have to be addressed in choosing an appropriate strategy for educational investment form the subject of this book. These include both broad and specific issues that must be tackled if resources are to be used as efficiently as possible, if the goals of equity are to be well served, and if educational investment is to be coordinated with other forms of investment to have the maximum effect on the development of human resources, the alleviation of poverty, and the growth of income and employment.

The book draws both on World Bank analysis and research and on the wider literature on planning educational investment. Although lessons are drawn from World Bank experience with education projects, the book is not primarily concerned with the detailed preparation, design, and implementation or evaluation of projects, but rather with the wider issues of analyzing educational investment. It is mainly for those who are responsible for formulating policy on investment in human development in developing countries.

Considerable attention is therefore given to the many questions that must be dealt with in designing such policies and in choosing between alternative ways of using scarce resources. The principal aim of the book, however, is to make planners and policymakers more aware of the problems that can arise in developing investment strategies for education and the analytical tools and information that are available to help solve them. Those who must implement the policies and projects will want to refer to the more specialized literature cited in the book, as will those who need more detailed information on the techniques discussed here.

The unifying theme of the book is the analytical and research experience of the Bank in this area and the lessons derived from operational experience in lending for education. In fact, one of the book's most useful functions is that it reviews the results of some of the lesser known or unpublished studies carried out at the Bank. At the same time, it emphasizes that more information will be needed if education is to become even more effective in promoting development.

AKLILU HABTE
Director
Education and Training Department
The World Bank

May 1985

Contents

Education for Development

An Analysis of Investment Choices

1

Introduction

Education, like other forms of investment in human capital, can contribute to economic development and raise the incomes of the poor just as much as investment in physical capital, such as transport, communications, power, or irrigation. The World Bank, which provides financial and technical help for the development of poor countries, has long recognized the importance of investment in education and has been active in this field since 1962.

When the Bank was first established in 1944, however, education was not counted among the productive purposes for which it was authorized to provide investment capital. This attitude prevailed right through the 1950s, when official statements of Bank policy took the view that the Bank should concentrate its lending on projects designed to make a direct contribution to the productive capacity of its members, rather than finance projects such as the construction and equipment of schools, colleges and universities. Even though such projects were considered essential to the development of a country, the Bank's proper role, as a bank, was said to be confined to providing loans for the more directly productive sectors of the economy. Gradually, this attitude changed, however, both within the World Bank and in the world at large.

One of the principal reasons for the shift in thinking was the growing interest during the 1960s in the economic value of education, which was described by Bowman (1966) as the "human investment revolution in economic thought." Before long, economists were trying to measure the contribution of education to economic growth (Schultz 1961, 1963; Denison 1962, 1967; Krueger 1968), and many were examining the concept of investment in human capital (Becker 1964, 1975). This activity influenced governments, planners, international agencies, and educators throughout the world. It also helped to fuel the growing demand for education, which had been sparked by the political independence recently gained in many of the developing countries.

Eventually World Bank policy reflected this recognition that education is a productive investment in human capital. The International Development Association (IDA) was established as a Bank affiliate in 1960 to

3

accelerate economic development and promote higher standards of living and economic and social progress in the poorest countries. Education was then seen as one of the most important ways of contributing to social progress, and in 1962 the first education project was initiated. The justification for this investment was that education is not only a basic human right, but also a basic component of social and economic development, and that properly planned investments in education pay great economic dividends, especially in the poorest countries.

Today, few would dispute that human development contributes to economic growth no less than physical capital. Yet many developing countries have become disillusioned about the economic value of education because the skill shortages of the 1960s have been replaced by growing unemployment among the educated, and poverty and enormous disparities of wealth and opportunities persist, despite massive expenditures on education. Thus, several vital questions with respect to education remain to be answered: How, for example, should governments plan investment in education in the face of growing financial constraints and the undiminished thirst for educational opportunities? How can scarce resources be used as productively and efficiently as possible? And, to what extent can education contribute to the goals of economic growth and social equity? Some of the answers can no doubt be found in the lessons of past experience.

Since 1962 the World Bank has invested more than $5 billion in about 260 education projects in more than 90 countries. This investment has helped provide a total of 3 million school, college, or university places, in more than 20,000 institutions. World Bank experience in the field of educational investment is therefore substantial. Moreover, this investment has been accompanied by assessment and careful analysis of implementation problems and achievements. In addition, the World Bank has commissioned research on the contribution of education to human and economic development, and has cooperated both with other international agencies and with individual countries in carrying out research on the economic value of education and disseminating the results.

World Bank Experience with Educational Investment

Since the World Bank first began to finance educational investment in 1962, it has contributed to a substantial increase in educational provision in developing countries, to an improvement in the geographical and social distribution of education opportunities, and to a raising of standards and quality of education in many countries. It has contributed, also, to the search for more efficient ways of using scarce resources; for example, through the use of educational technology, and through both its lending and training activities, the Bank has tried to promote better planning of educational investment. The purpose of this book is to draw on this rich experience of educational development, to summarize the practical lessons that have been learned, and to examine the educational investment issues of concern to policymakers and planners in developing countries.

There have been a number of important changes in emphasis in World Bank education projects since 1962. Initially, lending for education could not be for primary education or for liberal arts colleges, but had to be restricted to engineering, technical, managerial or other vocational education closely allied with other Bank projects. Furthermore, investment was confined to bricks and mortar. Today, however, the Bank is investing in primary and basic education (including nonformal education) as well as in technical and vocational education and teacher training. It is also investing in curricula reform, school textbooks, and other software as well as school buildings and equipment. Thus the emphasis now is on both qualitative and quantitative improvements, and there is as much concern with the equity as with the efficiency of educational investment.

The World Bank's present strategy in investing in education can be summarized by the following principles:

- Basic education should be provided for all children and adults as soon as the available resources and conditions permit. In the long term, a comprehensive system of formal and nonformal education should be developed at all levels.
- To increase productivity and promote social equity, educational opportunities should be provided without distinction of sex, ethnic background, or social and economic status.
- Education systems should try to achieve maximum internal efficiency in the management, allocation, and use of available resources so as to increase the quantity and improve the quality of education.
- Education should be related to work and environment in order to improve, quantitatively and qualitatively, the knowledge and skills necessary for economic, social, and other development.
- To satisfy these objectives, developing countries will need to build and maintain their institutional capacities to design, analyze, manage, and evaluate programs for education and training (World Bank 1980, p. 10).

Within this broad framework, there is scope for considerable diversity, as can be seen in the types of projects that have been funded, which range from rural primary schools to university faculties of engineering and technical institutes. Primary and nonformal basic education, including literacy programs, have received increasing support from the Bank in recent years, and more emphasis has been placed on improving access of the rural and urban poor to education. Although questions of equity received far more attention during the 1970s than in the first decade of World Bank involvement, both internal and external efficiency have always been emphasized in its education projects.

The shift from an overriding concern with bricks and mortar to an emphasis on school textbooks and curriculum development has been accompanied by a greater interest in the planning, administration, and management of education programs, and on providing technical assistance for educational evaluation. Much of the Bank's investment in education projects is therefore concerned with institution building, and many projects are implemented by units that can influence the wider institutional structure. Many Bank-financed projects have had a "multiplier" effect that goes well beyond the limited scope of the project itself. The success of the Bank's efforts in the education field is not measured primarily by the amount of money lent, but rather by the effectiveness with which Bank and country resources are used to meet crucial needs. Thus, "the motivating and multiplying effects of external assistance will be the principal test of success" (World Bank 1972, p. 230).

Bank-financed education projects are large-scale investments (the average loan size in 1983 was $26 million) and they have long gestation periods (usually six to eight years from identification to completion). It takes even longer to judge whether a project has been successful. The effects of education on employment, for example, or the external efficiency of an education project cannot be properly assessed for many years. As a result, more and more education projects now include evaluation procedures, including tracer studies, which follow samples of students from schools to employment. Such procedures may make it possible to discover how education contributes to manpower and employment goals.

Another point to note is that World Bank assistance for education is usually project oriented. The specific activities within a project are planned and agreed upon in advance, and funds are earmarked to cover their cost. This means that a project must be fully defined and prepared before the Bank can assist in financing it. The Bank has recently begun to consider the provision, under appropriate circumstances, of "sector loans," which would focus attention on broader issues and would transfer responsibility for project identification and supervision from the Bank to the borrower. This step is one indication of the increasing attention being given to borrower involvement, particularly government participation in identifying and preparing projects. The Bank now recognizes that an important objective of educational lending should be to assist borrowers to become permanently equipped for the efficient development, management, researching and evaluation of their national education and training systems.

One of the present criteria for Bank assistance in education is that the government must demonstrate its ability to meet the costs of operating project institutions after the completion of a project. The Bank has found that the largest single constraint on the volume of its lending for education has been that because of existing commitments for current expenditure, countries are unable to meet the operating costs generated by a project. The choice of projects for Bank assistance is therefore heavily influenced by the capacity of governments in developing countries to meet the recurrent costs of the projects. Generally, these must be borne by the government. The Bank believes that if a government knows that it will be required to pay recurrent costs, its involvement in all aspects of the project, and its commitment to it, will be all the greater. However, *initial* operating costs may be covered by Bank lending to ensure that a project gets off to a good start, but only if the government demonstrates its ability to assume responsibility for them after a year or two, within the life of a project.

Every Bank-assisted project must contribute substantially to development objectives and be economically, technically, and financially sound. All projects pass through a series of stages, known as the "project cycle" (Baum 1982):

1. *Identification* of projects with a high priority, which fit into and support a country's overall national and sectoral policies and objectives.
2. *Preparation* of a project, which includes full specification of its objectives; the timetable for achieving them; the technical, institutional, economic, and financial conditions necessary for success; and the relative costs and effectiveness of alternative ways of achieving the project's objectives.
3. *Appraisal* of the project, in terms of its technical, institutional, economic, and financial feasibility. This appraisal is used to establish the terms and conditions of a loan.
4. *Negotiation* between the Bank and the borrower concerning the points to be set out in the loan documents—namely, the terms of the loan, the legal obligations of both Bank and borrower, and all the conditions that have to be fulfilled.
5. *Implementation and supervision* of the project. If implementation problems arise, as often happens, the project may have to be modified.
6. *Evaluation* of the project, which constitutes the final stage of the cycle. At completion, all projects are audited by the Bank's Operations Evaluation Department (OED) and a project completion

report is prepared by the regular project staff or the borrower. Thus, every project is subjected to both self-evaluation and independent audit by OED; in addition, some projects are selected for extensive long-term evaluation, which is an assessment of their impact at least five years after final disbursement.

The aim of this project cycle is to ensure that the lessons of experience are built into the design and preparation of future projects.

The Education Project Cycle

Before the cycle of an education project can begin—that is, before the project can even be identified—considerable preliminary work must take place. First, the extent of government interest in securing Bank assistance for education must be established, and an education sector study may be required to analyze the country's education and training system; to identify its objectives, strengths, and weaknesses; and to suggest alternative strategies for the development of the sector and broad investment priorities.

There are no hard and fast rules concerning investment priorities since they must reflect the individual needs and conditions of the country. The Bank recognizes that countries may require assistance in many different areas, and that the choice of investment priorities will depend on a country's specific situation. The evolution of the Bank's education sector policy, as noted earlier, reflects the gradual liberalization of the initial conditions for lending, which conform with the diversity of educational and development needs. An education sector study for a country will try to identify priorities or programs that merit investment assistance.

The next stage is known as project identification. The objective here is to identify a "package" of facilities and activities that would be eligible for Bank assistance. This stage frequently involves cooperation between the Bank and other international agencies such as Unesco, the International Labour Office (ILO), the United Nations Development Program (UNDP), or the Food and Agriculture Organization of the United Nations (FAO).

After a project has been identified, the detailed preparation begins; the number and location of project institutions are specified, together with the furnishing and equipment, staff training and technical assistance requirements, and their costs. In addition, detailed financial and administrative arrangements for project implementations are worked out, usually by the borrower, although the Bank may provide technical assistance at this stage. However, project appraisal is always carried out by Bank staff to establish that

- The content and objectives of the project are clearly defined and in line with the borrower's educational and development objectives.
- The country will be able to meet the capital and recurrent costs and staffing needs of the project.
- Project items meet Bank criteria and policies for education financing.
- The institutional arrangements for project management are satisfactory.

Considerable time elapses between education sector analysis and agreement on the terms of the loan for an education project. The average length of time between project identification and loan or credit authorization is two and a half years, which is longer than it takes to reach agreement in most other sectors of Bank operations. The fact that attempts to compress this time period are seldom successful reflects the complexity of the issues that have to be tackled before an education project is sufficiently advanced for Bank financing to be agreed to and implementation to begin.

The implementation period lasts even longer—on the average, seven years. During this time, projects frequently experience implementation problems, which may lead to cost overruns or may make it necessary to modify the design. Most delays are caused by inadequate project preparation, borrowers' difficulties with Bank procurement procedures, or specific problems connected with funding, social unrest, or natural disasters. Despite such problems, many new educational programs have been established, new educational technologies have been developed, and important educational innovations have been achieved. Now and then, innovative or experimental schools or programs run into problems, however, sometimes because the difficulties of innovation have been underestimated, or the borrower's commitment to the innovations has been less than was hoped.

The Bank's education projects cannot be completely evaluated because it is not yet possible to measure the long-term effects of educational investment. Nevertheless, after twenty years of involvement in educational investment—during which the World Bank has amassed a wealth of practical experience and a substantial volume of research on the economic contribution of education—some general conclusions can be drawn about the success of World Bank activity in the field of educational investment.

Not only has the World Bank's involvement in education and training grown rapidly, but it has also attracted additional investment to the education sector, and thereby has helped to expand and improve education in developing countries. As a result of the Bank's ongoing support, these countries have been able to increase the efficiency of their education system, for example, through more intensive use of school facilities and improved school building design, and to increase the supply of skilled manpower. According to staff evaluations, most Bank projects have been in line with borrowers' needs and priorities, have contributed to the development of new attitudes toward education, and have helped to strengthen administrative and planning skills. Some have brought education to the less privileged groups in society, while others have improved the internal efficiency of schools and other institutions; still others have helped to upgrade science teaching and have introduced more practical work in technical, vocational, and agricultural education.

Thus, World Bank financing has helped countries to expand education enrollment, improve access to education, and increase its quality and relevance. The investment has proved worthwhile, insofar as this can be measured at the present time. Even so, some enormous problems have yet to be resolved. Access to education remains uneven, the quality of education is often poor, and many countries cannot yet afford to provide even basic education for all who want the opportunity. Moreover, the recurrent costs of education are constraining further investment. Meanwhile high rates of wastage, including both repetition and dropout, hint at internal inefficiencies in education, and despite skilled manpower shortages in some areas, unemployment among school leavers or college or university graduates is a growing problem in many countries.

If educational investment is to help solve these problems, the choice between alternative investment priorities and the identification and preparation of education projects must be economically sound, as well as technically and financially viable.

The Economic Analysis of Educational Investments

The purpose of this book is to show, through research and experience, how economic analysis of investment choices may help determine investment priorities for education and help in the identification and preparation of education projects. The main concern here is with planning education investment, rather than with the details of project design and implementation, the technical aspects of procurement, or Bank lending procedures. Thus it should be of interest to those in developing countries who are responsible for formulating general policy on investment in human development, rather than those involved in detailed project preparation and implementation, such as architects, surveyors, or education specialists.

The following chapters examine issues that must be tackled in assessing the need for investment in education, evaluating its contribution to economic development, and choosing between alternative investment projects. World Bank experience has not yielded ready-made solutions to the complex problems of choosing the most efficient and productive types of educational investment. On the contrary, experience has shown that solutions to many problems—such as the level of education likely to offer the highest returns, the right balance between technical and general education, and the combinations of inputs and technologies likely to produce the greatest output of skills and knowledge and be most effective in preparing people for employment—depend in some degree on a country's economic and social conditions, physical and human endowments, and political priorities.

The economic techniques that have been offered as guides to educational investment decisions—cost-benefit analysis, manpower analysis and forecasting, cost-effectiveness analysis—all have a part to play in making the choice, but none is sufficient by itself. Nor will economic analysis of investment projects, by itself, show how educational investment may contribute most effectively to social development. Experience repeatedly demonstrates the need for an interdisciplinary approach to the analysis of education and other forms of investment in human capital. For this reason, World Bank studies on education have included research not only by economists, but also by sociologists, communications experts, statisticians, educational psychologists, and other educational specialists. This book draws on their findings and on the wider literature, though not exhaustively. Although the book deals primarily with the economic evaluation of educational investment, it recognizes that unless political and social factors are taken into account, educational investment cannot make its full contribution to development.

Outline of the Book

The discussion opens in chapter 2 with the general criteria for choosing between alternative investments, particularly the economic criteria of costs and benefits, costs and effectiveness, and the contribution of education to economic growth. The problems in alleviating poverty in developing countries are also considered. These criteria are certainly not the only ones relevant to educational investment issues. Nevertheless, the fact that Bank loans and IDA credits are primarily designed to promote economic development by providing financial support for sound, productive projects means that economic criteria are bound to play an important role in the choice between alternative projects.

The remainder of the book is devoted to a detailed examination of these economic criteria. Chapter 3 considers the role of cost-benefit analysis in measuring the profitability of educational investment, the practical problems of measuring costs and benefits, and the theoretical controversies that still surround estimates of the rate of return to education. The topic of discussion in chapter 4 is the contribution of education to meeting the needs for skilled manpower. Of interest here is past experience in forecasting manpower requirements or demand, and in manpower analysis and planning.

Chapter 5 considers the factors that influence the private demand for education, while chapter 6 looks at the issues involved in financing education, including the role of public subsidies for education and the use of fees, student loans, and other methods of cost recovery.

Chapter 7 deals with costs and expenditure on education, particularly the patterns and determinants of costs, and suggests some ways of reducing unit costs, for example, through economies of scale.

In chapter 8, the discussion focuses on the internal efficiency of education, the use of cost-effectiveness analysis, and the various attempts that have been made to increase the internal efficiency of educational institutions in order to achieve greater output from available resources.

Chapter 9 takes up the important questions of equity and the inequalities of access and distribution of educational opportunities in developing countries. The effect of education on income distribution is also considered, together with the equity implications of alternative ways of financing education. Chapter 10 then looks at the links between educational investment and other sectors, including health and fertility, and agricultural and rural development. The discussion closes in chapter 11 with a summary of the main lessons to be drawn from twenty years of World Bank experience with investment in education.

References

Baum, Warren C. 1982. *The Project Cycle*. Washington, D.C.: World Bank.

Becker, Gary S. 1974. 2d ed. 1975. *Human Capital: A Theoretical and Empirical Analysis, with Special Reference to Education*. New York: National Bureau of Economic Research.

Bowman, M. J. 1966. The Human Investment Revolution in Economic Thought. *Sociology of Education* (Spring):111–37.

Denison, E. F. 1962. *The Sources of Economic Growth in the United States and the Alternatives Before Us*. New York: Committee for Economic Development.

———. 1967. *Why Growth Rates Differ: Postwar Experience in Nine Western Countries*. Washington, D.C.: Brookings Institution.

Krueger, A. O. 1968. Factor Endowments and Per Capita Income Differences among Countries. *Economic Journal* 78 (September):641–59.

Schultz, T. W. 1961. Education and Economic Growth. In *Social Forces Influencing American Education*, ed. N. B. Henry. Chicago: National Society for the Study of Education, University of Chicago Press.

———. 1963. *The Economic Value of Education*. New York: Columbia University Press.

World Bank. 1972. *World Bank Operations: Sectoral Programs and Policies*. Baltimore, Md.: Johns Hopkins University Press.

———. 1980. *Education*. Sector Policy Paper. Washington, D.C., April.

2

Criteria for Educational Investment

The basic economic problem all governments face is how to allocate scarce resources between competing ends. The resources consist of labor (which can be classified according to its different capacities, skills, and knowledge), capital, and land and other natural resources; each of these categories can be subdivided according to levels of quality. The competing ends are consumption (that is, the production of consumer goods and services) and investment. The choice between current consumption, which brings immediate satisfaction of needs and wants, and investment, which creates the capacity to produce future goods and services, is a matter of time preference, and it rests on political as well as economic considerations. Thus, it can be said to depend on society's objectives. The choice between alternative investments—for example, between investment in manufacturing industry and agriculture, or between factories, bridges, ships, or schools—also depends on society's objectives, and on the balance between the costs of the investment, in terms of scarce resources, and the future benefit to be derived from the investment.

Although the objectives of society are many and varied and not always well defined, the economic policies and choices of government are typically concerned with three objectives: satisfying immediate needs and wants through current production of goods and services, that is, consumption; increasing the supply of goods and services in the future through increased national income, that is, economic growth; and ensuring an adequate distribution of goods and services between different groups in society, that is, equity.

Different societies attach different priorities to these objectives, among which economic growth and equity are of prime concern to the World Bank. Other objectives—for example, the desire to create employment or control inflation—may also influence economic policy and a government's perception of the benefits of investment.

Every investment choice, then, is preceded by a number of fundamental decisions pertaining to these objectives. That is to say, governments first have to decide how scarce resources should be allocated between current consumption and investment in future wealth, what relative priorities should be given to growth and equity objectives, and what tradeoffs between these two objectives they are willing to accept. These decisions in turn influence evaluations of alternative investment opportunities in terms of their possible contribution to current consumption and future growth and their distributional impact.

Education represents both consumption and investment. On one hand, it is valued for its immediate benefits, but, on the other, it helps to create income in the future by providing educated workers with skills and knowledge that enable them to increase their productive capacities and thus receive higher earnings. This means that the distribution of education influences future income distribution, and thus that the equity implications of educational investment are extremely important. As chapter 1 has shown, education has not always been looked upon in this way. In the early 1960s it was considered a "basic human right" associated primarily with consumption. However, its value as an investment that can increase future productive capacity came to be recognized, so that now both ends are taken into account when priorities are being established for the allocation of funds. In determining the balance between investment in human capital (particularly education and training) and investment in physical capital and infrastructure, policymakers must consider two important questions:

- Does education contribute to economic growth?
- How does the contribution of education to economic development compare with the contribution of physical capital?

The Contribution of Education to Economic Growth

The concept that investment in human capital promotes economic growth actually dates back to the time of Adam Smith and the early classical economists, who emphasized the importance of investing in human skills. In the 1960s Schultz (1961) and Denison (1962) showed that education contributes directly to the growth of national income by improving the skills and productive capacities of the labor force. This important finding led to a flood of studies on the economic value of investment in education. Research in this area slowed in the 1970s, however, because of a lack of economic growth and a certain ambivalence about the role of education in development. Recently, the World Bank has expressed renewed interest in human development, particularly education, as is reflected in its 1980 *World Development Report*. Drawing on research by Hicks (1980) and Wheeler (1980), this report reaffirmed the importance of education in promoting economic growth. The contribution of education to growth is even stronger if the complementarities between education and other forms of investment are taken into account (Psacharopoulos 1984).

The early attempts to measure the contribution of education to economic growth were based either on the growth accounting approach, used by Denison and others, or on the rate of return to human capital, an approach adopted by Schultz and others. Growth accounting is based on the concept of an aggregate production function, which links output (Y) to the input of physical capital (K) and labor (L). The simplest form of production function, assumed in many of these studies, is a linearly homogeneous production function: $Y = F(K, L)$.[1]

If economic growth is due entirely to increases in physical capital and labor, then it should be possible to disaggregate the rate of growth of output into its capital and labor components. Denison's attempt to explain U.S. economic growth between 1910 and 1960 in terms of increases in labor and physical capital immediately established, however, that there was a large "residual" that could not be explained in this way. This posed a challenge for researchers, who then turned their efforts to discovering how much of this "residual" was related to the effect of education on the quality of the labor force, and how much to other factors such as improvements in the quality of physical capital and economies of scale. In this respect, Denison calculated that between 1930 and 1960, for example, almost a quarter (23 percent) of the rate of growth of output in the United States was due to the increased education of the labor force.

When he made a similar calculation for the period since 1950, Denison (1967) found the contribution of education in the United States to be only 15 percent, while that in other advanced countries varied considerably, from 2 percent in Germany, to 12 percent in the United Kingdom, 14 percent in Belgium, and 25 percent in Canada. When the same approach was applied to the growth rates of some developing countries (Nadiri 1972), the results were mixed. Here, education appeared to account for 16 percent of the increase in output in Argentina, but less than 1 percent in Mexico, and only 2 to 3 percent in Brazil and Venezuela (see table 2-1).

Schultz's (1963) method of measuring the contribution of education to economic growth (that is, in terms of the rate of return to human capital, which he then compared with the rate of return to physical capital) led him to suggest, as Denison had, that a substantial proportion of the rate of growth of output in the United States was due to investment in

Table 2-1. *The Contribution of Education to Economic Growth*

Country	Percentage contribution to annual growth rate
North America	
Canada	25.0
United States	15.0
Europe	
Belgium	14.0
Denmark	4.0
France	6.0
Germany, Fed. Rep.	2.0
Greece	3.0
Israel	4.7
Italy	7.0
Netherlands	5.0
Norway	7.0
United Kingdom	12.0
U.S.S.R.	6.7
Latin America	
Argentina	16.5
Brazil	3.3
Chile	4.5
Colombia	4.1
Ecuador	4.9
Honduras	6.5
Mexico	0.8
Peru	2.5
Venezuela	2.4
Asia	
Korea, Rep. of	15.9
Japan	3.3
Malaysia	14.7
Philippines	10.5
Africa	
Ghana	23.2
Kenya	12.4
Nigeria	16.0

Source: Psacharopoulos (1984), p. 337.

education.[2] This method has also been used to estimate the contribution of education to economic growth in developing countries, including Ghana, Kenya, Nigeria, Malaysia, and the Republic of Korea. (The estimates presented in table 2-1 for developed and developing countries are based on the methodologies used by both Denison and Schultz.)

The overall conclusion is clear: increased education of the labor force

Table 2-2. *Economic Growth and Life Expectancy, Selected Economies*

Economy	Growth rate,[a] 1960–77 (percent)	Life expectancy, 1960 (years)	Deviations from expected levels of life expectancy (years)	Adult literacy, 1960 (percent)	Deviations from expected levels literacy, 1960 (percent)
Singapore	7.7	64.0	3.1	n.a.	n.a.
Korea, Rep. of	7.6	54.0	11.1	71.0	43.6
Taiwan	6.5	64.0	15.5	54.0	14.2
Hong Kong	6.3	65.0	6.5	70.0	6.4
Greece	6.1	68.0	5.7	81.0	7.5
Portugal	5.7	62.0	4.7	62.0	1.7
Spain	5.3	68.0	1.8	87.0	1.2
Yugoslavia	5.2	62.0	4.7	77.0	16.7
Brazil	4.9	57.0	3.0	61.0	8.6
Israel	4.6	69.0	2.0	n.a.	n.a.
Thailand	4.5	51.0	9.5	68.0	43.5
Tunisia	4.3	48.0	−0.5	16.0	−23.8
Average: top 12	5.7	61.0	5.6	64.7	12.0
Average: all	2.4	48.0	0.0	37.6	0.0

n.a. Not available.

a. Growth rate of real per capita GNP.

Source: Hicks (1980), p. 12.

appears to explain a substantial part of the growth of output in both developed and developing countries since about 1950. However, these estimates rest on a wide variety of theoretical assumptions that have been challenged. In particular, it is assumed that the earnings of different groups of workers are a measure of their contribution to output; that the higher earnings of educated workers are a measure of their increased productivity, and therefore of their contribution to economic growth; and that the relationship between inputs and output is a fairly simple one, which can be analyzed in terms of an aggregate production function.

All these assumptions have been attacked in the literature on the economics of education and this opposition has helped to undermine the widespread belief that investment in education contributes to economic growth. Recent attempts to use econometric techniques to relate inputs to output, however, have again demonstrated the link between education and growth of output. Recent research for the World Bank, for example, provides evidence of the link between various aspects of human resource development and economic growth. One such study (Hicks 1980) examined the relationship between growth and literacy, as a measure of educational development, and life expectancy in eighty-three developing countries during the period 1960–77 and found that the twelve developing countries with the fastest growth rate had well above average levels of literacy and life expectancy (table 2-2). According to these results, not only do literacy levels rise with the level of national income, but these twelve countries have higher levels of literacy and life expectancy than would be predicted for countries of that income level on the basis of the regression between literacy and per capita income. In the case of Korea and Thailand, the considerable difference between actual and "expected" literacy levels suggests that rapidly growing countries have well-developed human resources. It does not, of course, prove the opposite: that countries with high levels of human resource development will thereby achieve faster economic growth. Hicks did find, however, that the twelve developing countries with the highest level of life expectancy had high rates of economic growth.

Further analysis by Hicks confirms the existence of a relationship between economic growth and human resource development, as measured by literacy and life expectancy. Three variables were found to explain about 60 percent of the variation in per capita growth rates in developing countries between 1960 and 1977: the rate of investment, the growth rate of imports, and the level of human resource development in 1960.

Of course, correlation does not prove causation. This objection automatically weakens any argument that uses regression analysis to prove that educational development causes economic growth. Early work by

Bowman and Anderson (1963), Kaser (1966), and others certainly demonstrated a correlation between level of per capita income and level of educational development. However, the fact that rich countries have higher levels of literacy and spend more on education than poor countries could mean that education helps countries become rich, or it could mean that rich countries can afford to spend more on education.

To allow for the fact that education, and other indicators of human resource development, are both a result of and a cause of economic development, Wheeler (1980) devised a simultaneous model, which he applied to data for eighty-eight developing countries. The simultaneous model takes into account the interactions, over time, between growth and human resource development, and tries to separate cause and effect. Tests with this model suggest that education, health, and nutrition contribute to growth of output not only directly, but also indirectly, by increasing the rate of investment and by lowering the birth rate. The relationship is stronger for education than for life expectancy, however, which shows a strong relationship with economic growth in the 1960s, but not in the 1970s. Wheeler found that on the average an increase in the literacy rate from 20 to 30 percent causes national income (GDP) to increase by 8 to 16 percent; he also found that the relationship is even stronger in African countries. After examining data for sixty-six developing countries in a similar analysis, Marris (1982) concluded not only that education strongly affects economic growth, but that general investment has less effect on growth rates when it is not supported by educational investment.

Other research, too, has demonstrated that investment in education complements investment in physical capital. For example, World Bank research on the links between education and the productivity of farmers (Jamison and Lau 1982) has shown that investment in improved seeds, irrigation, and fertilizers is more productive, in terms of increased crop yields, when farmers have four years of primary education rather than none.

Thus it appears that previous attempts to measure the contribution of education to economic growth may have underestimated the effects of education by ignoring its indirect effects. There are other grounds, too, for thinking that previous attempts to measure the economic contribution of education may have underestimated the effect of education on growth. Both Denison and Schultz, for example, concentrated on formal education and ignored the contribution of informal education and on-the-job training. As Cochrane (1979) has pointed out, education also has important links with other aspects of human resource development, notably health and fertility (see chapter 10).

There is ample evidence that education makes both a direct and an

indirect contribution to economic growth, but the chicken-and-egg relationship between education and growth can never be fully established.
Nonetheless, strong support can be found for the notion that the most
likely causal link is from education to economic growth, rather than the
other way around. For example, when Easterlin (1981) examined the
relationship between education and economic growth in twenty-five of
the largest countries in the world, he concluded that the spread of the
technology of modern economic growth depended on the greater learning potential and motivation arising from the development of formal
schooling. Thus, economic history, together with recent econometric
research, confirms the belief that investment in education can contribute
to economic growth.

The question of whether it is more profitable to invest in men or
machines cannot be answered simply. Some evidence (see figure 2-1)

Figure 2-1. *The Social Rate of Return to Physical and Human Capital,
by Level of Economic Development*
(intercountry averages)

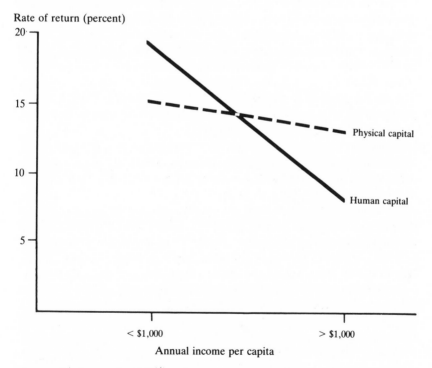

Source: Psacharopoulos (1973), p. 8.

suggests that in developing countries the average rate of return to human capital is higher than the rate of return to physical capital, whereas in more developed countries, the reverse is true. It is also true, however, as has already been pointed out, that investment in education often complements investment in physical capital and makes it more productive.

Although none of this research shows which types or levels of education are likely to have the greatest impact on productivity or growth, it does show that spending on education should be regarded as a productive investment, rather than pure consumption. If governments wish to maximize growth, however, they need to know how educational investment compares with other forms of investment—particularly investment in physical capital and social infrastructure—and which forms of educational investment offer the highest returns. In other words, educational investment must be evaluated in terms of opportunity costs and the relationship between costs and expected benefits.

The Opportunity Cost of Educational Investment

Since resources are limited, some opportunities have to be sacrificed when investment decisions are finally made. These lost opportunities can be regarded as part of the cost of the investment, which is thus said to include *opportunity costs*. If resources are invested in education, for example, they are no longer available for investment in health, industry, or agriculture. When a decision is made to build a new school, consideration should be given to the fact that alternative opportunities to build a dam, a road, or a factory or to provide basic health care must be sacrificed.

The opportunity cost of any project consists of the value of all resources in their next best alternative use, and this cost is wider than the financial cost of the project, measured in money terms. In particular, resources such as the time of students and pupils, the goods and services provided directly by the local community (for example, free food or housing for teachers), or voluntary labor to build a new school must all be counted as part of the opportunity cost of investment in education, even if no financial cost is involved. The point is not whether resources have a financial cost, but whether they have an alternative use.

Thus the value of a student's time may be measured in terms of forgone earnings or agricultural product, that is, the value of the alternative opportunity—productive employment—that both the student and society must sacrifice when the former spends his or her full time in being educated, rather than working. If unemployment is such that the alternative opportunities for productive work are reduced, the opportunity cost of education will also be reduced. But even when unemployment is high, the opportunity cost of students' time is seldom zero. Indeed, in developing countries the opportunity cost of time spent on education by pupils in primary school may be quite high, particularly for poor families, since even young children may contribute to family income by working in the fields, carrying water, or looking after babies, and thus can free adults for more productive work. (The difference between opportunity cost and money expenditure is examined in more detail in chapters 3 and 7, which also describe the use of shadow prices to measure the economic value of resources when distortions in market prices mean that expenditure does not reflect the true value of goods or services.)

The opportunity cost must be considered in the evaluation of investment projects because, as noted earlier, every investment decision involves a sacrifice of alternative opportunities. For governments, or for the World Bank, the justification for any investment must be that it will make the greatest possible contribution to society's objectives, as currently perceived. If it does not, then the limited resources are not being allocated as efficiently as possible. The choice of investments must, therefore, be based on an analysis of the *external efficiency* of all competing uses of resources, from the point of view of society's objectives, as well as the *internal efficiency* of resource use. Both internal and external efficiency must be at a maximum level if the best use is to be made of scarce resources. This means that investment choices must be based both on cost-benefit analysis, which is concerned with external efficiency, and on cost-effectiveness analysis, which measures internal efficiency.

For example, the choice between investing in more primary- or university-level education should be based on both costs and benefits. The opportunity cost of investing in a technical university may be that the country cannot afford thousands of potential primary school places. In industrialized countries, the cost of a university education may be double or treble the cost of elementary education, but in some developing countries, the cost difference may be as great as 50:1 or even 100:1. This high opportunity cost of university-level education must be assessed in relation to the expected benefits of the two types of investments before it will be possible to justify devoting a large share of society's scarce resources to expanding university enrollment. Since the objectives of growth and equity may themselves compete, the opportunity cost and benefits of investment must be compared in terms of both increased future income and redistribution of income. (The possible tradeoffs between growth and equity objectives are discussed in more detail in chapter 9.)

Criteria for Evaluating Investment Projects

The following chapters are concerned with the techniques used to evaluate educational investment, including cost-benefit analysis, cost-effectiveness analysis, and the analysis of demand for and utilization of manpower. In many cases, decisions about educational investment have to be based on crude techniques because data on education are not always plentiful, particularly in developing countries. The techniques themselves—such as rate-of-return calculations, forecasts of manpower requirements, or analysis of economies of scale—are not as important as the criteria for evaluating alternative investment opportunities.

Although it may be impossible, for example, to measure the rate of return to investment in education with any precision, it is nevertheless important to use the framework of cost-benefit analysis to compare the costs and future income of alternative projects. Similarly, it is impossible to forecast accurately the future demand for skilled manpower of different categories, but it is important to examine the macroeconomic condition affecting future job opportunities for workers with different levels or types of education, and to look at current patterns of utilization in terms of pay levels and proportions of workers who are employed or unemployed. Nor is it possible to measure all the outputs of education (that is, the knowledge or skills transmitted, the creation or change of attitudes toward modernization, and the effects of these attitudes or skills on productive capacity and social welfare); yet, it is important to compare the costs and effectiveness of different types of education, curricula, or teaching methods. In sum, qualitative judgments are just as significant in investment decisions as quantitative evaluations.

World Bank experience with educational investment has led to a gradual widening of the criteria used for evaluating investments. In the early 1960s, the main criterion was the need for workers with technical or other vocational skills, and the goal was simply to increase the supply of skilled manpower. During the 1970s, however, the experience of developing countries, together with research on the economic contribution of education, showed that high-level manpower shortages were not the only obstacle to economic growth. Basic education was found to be important and in many cases to offer higher returns (that is, higher productivity of educated workers) in relation to the costs of education. Early studies of the economic returns to education concentrated on workers in the industrialized, urban sector, but subsequently attention shifted to the rural sector and to the contribution of both formal and nonformal education to farmer productivity. Thus, the World Bank's early education projects were designed to meet quantitative demand, and the emphasis was on satisfying enrollment targets. These concerns have now been widened to include quality and social and geographical distribution of enrollments.

The attention given to equity criteria has grown steadily since the first decade of World Bank involvement in educational investment, and the criteria themselves have undergone change. Initially, for example, some countries abolished tuition fees in an attempt to equalize educational opportunities and ensure that poor students were not denied access. Recently, however, the equity of existing policies on fees, student aid, and other subsidies has come into question as a result of studies on the distribution of the costs and benefits of education, the quality of schooling, and access. It is now recognized that the goal of equity in educational investment is tied up with complex qualitative and quantitative issues (these are explored in chapter 9).

All Bank-financed investment must satisfy four types of criteria: technical, institutional, economic, and financial. This book is primarily concerned with the economic evaluation of education investment, but the other criteria should not be neglected. Technical criteria would include design standards for schools, for example, and the feasibility of alternative educational technologies, such as educational television. Institutional criteria would be the impact of the project on the educational system as a whole and on its managerial capacities. Experience has shown, for example, that the effectiveness of curriculum reform or the introduction of a new type of school, such as the diversified secondary schools introduced in several developing countries, may be reduced if institutional rigidities are ignored, if teacher training is not adapted to new curriculum needs, or if existing examination procedures remain unchanged. Financial criteria take account of future recurrent and capital costs, requirements for foreign exchange as well as domestic currency, and unexpected contingencies. Financial viability with respect to prices and expected cash flow is also significant, particularly in the case of projects designed to produce revenue through the sale of goods and services. Cost recovery is also emphasized in some developing countries. (Chapter 6 looks at the possibilities for cost recovery in education, including tuition fees, charges for board and lodging in schools, colleges or universities, and financial aid for students in the form of student loans, rather than scholarships, grants, or bursaries.)

Economic evaluation of educational investment projects should take into account the following criteria:

- Direct economic returns to investment, in terms of the balance between the opportunity costs of resources and the expected future benefits measured by increases in the productivity of educated workers

- Indirect economic returns, in terms of external benefits affecting the incomes of other members of society
- Fiscal benefits in the form of higher taxes paid as a result of increased incomes
- Satisfaction of demand for skilled manpower, which takes into account pay patterns, employers' hiring practices, and other indicators of manpower utilization
- The private demand for education (in the light of the private rate of return to educational investment, the level of fees, and other private costs) and social and other factors determining individual demand for education
- Internal efficiency of educational institutions, in terms of the relationships between inputs and outputs, measured by wastage and repetition rates, and other cost-effectiveness indicators
- The geographical and social distribution of educational opportunities
- The distribution of financial benefits of education and financial burdens
- The effect of the distribution of educational opportunities on income distribution and the contribution of education to the reduction of poverty
- The links between educational investment and investment in other sectors, including health and agricultural development.

The remainder of this book is devoted to a detailed examination of these criteria as they relate to experience with educational investment and research. The discussion should throw some light on the fundamental questions that must be dealt with by all those responsible for planning educational investment: How should resources be allocated between education and other forms of social investment, including health and investment in physical capital? How should resources be allocated within education, between different levels, between general and specialized education, and between different types of school or different methods or techniques of teaching? There is now ample evidence that education can contribute to economic growth, as well as satisfy basic human needs. Unless such criteria are used to evaluate alternative investment strategies and projects, educational investment will not be able to make the maximum contribution to growth and equity, and resources will not be used as efficiently and effectively as possible. In other words, policymakers must recognize that all investment decisions rest on a *choice* between alternative opportunities with respect to costs and benefits, costs and effectiveness, and equity implications.

Notes to Chapter 2

1. In fact, Denison implicitly assumes a linearly homogeneous production function, known as the Cobb-Douglas production function. For further details of the method of growth accounting used by Denison, see Psacharopoulos (1984).

2. Theoretically, the method adopted by Schultz and by Denison would lead to the same conclusion, in many cases, but differences in the way inputs are measured lead to computational differences. For further discussion of the two approaches, see Psacharopoulos (1973), pp. 111–18.

References

Bowman, M. J., and C. A. Anderson. 1963. Concerning the Role of Education in Development. In *Old Societies and New States*, ed. C. Geertz. Glencoe, Ill.: Free Press.

Cochrane, Susan H. 1979. *Fertility and Education: What Do We Really Know?* Baltimore, Md.: Johns Hopkins University Press.

Denison, E. F. 1962. *The Sources of Economic Growth in the United States and the Alternatives Before Us*. New York: Committee for Economic Development.

———. 1967. *Why Growth Rates Differ: Postwar Experience in Nine Western Countries*. Washington, D.C.: Brookings Institution.

Easterlin, R. 1981. Why Isn't the Whole World Developed? *Journal of Economic History* 41 (March):1–19.

Hicks, Norman. 1980. *Economic Growth and Human Resources*, World Bank Staff Working Paper no. 408. Washington, D.C.

Jamison, David T., and Laurence J. Lau. 1982. *Farmer Education and Farm Efficiency*. Baltimore, Md.: Johns Hopkins University Press.

Kaser, M. 1966. Education and Economic Progress: Experience in Industrialized Market Economies. In *The Economics of Education*, ed. E. A. G. Robinson and J. Vaizey. Proceedings of a Conference of the International Economic Association. New York: St. Martin's Press.

Marris, R. 1982. Economic Growth in Cross Section. London: Birkbeck College, Department of Economics. Processed.

Nadiri, M. I. 1972. International Studies of Total Factor Productivity: A Brief Survey. *Review of Income and Wealth* 18, no. 2 (June):129–54.

Psacharopoulos, George. 1973. *Returns to Education: An International Comparison*. Amsterdam and New York: Elsevier and Jossey-Bass.

———. 1984. The Contribution of Education to Economic Growth: International Comparisons. In *International Productivity Comparisons and the Causes of the Slowdown*, ed. J. Kendrick. Cambridge, Mass.: Ballinger.

Schultz, T. W. 1961. Education and Economic Growth. In *Social Forces Influenc-ing American Education*, ed. N. B. Henry. Chicago: National Society for the Study of Education, University of Chicago Press.

———. 1963. *The Economic Value of Education*. New York: Columbia Univer-sity Press.

Wheeler, D. 1980. *Human Resource Development and Economic Growth in Developing Countries: A Simultaneous Model*. World Bank Staff Working Paper no. 407. Washington, D.C.

World Bank. 1980. *World Development Report 1980*. New York: Oxford Univer-sity Press.

3

Cost-Benefit Analysis of Educational Investment

For both governments and individuals, the choice between different ways of investing resources rests to a great extent on an evaluation of the costs and benefits associated with the investments. The alternatives will differ as to the magnitude of the costs that must be incurred, the expected benefits that will be generated, the time scale of both costs and benefits, and the uncertainty or risks surrounding the project. Cost-benefit analysis is a technique by which these factors can be compared systematically for the purpose of evaluating the profitability of any proposed investment.

An investment is considered a profitable use of resources for the individual or for society as a whole when the expected benefits exceed its costs. Thus, in choosing between alternative investments, individuals or governments try to evaluate both costs and benefits and identify the investments that will achieve the greatest possible benefit in relation to cost.

The technique of cost-benefit analysis has been developed to make this evaluation as systematic, reliable, and comprehensive as possible and to eliminate the need for guesswork, hunch, or intuition. Cost-benefit analysis is an aid to judgment, however, not a substitute for it, since future costs and benefits can never be predicted with certainty, and measurement, particularly with respect to the likely benefits of a project, can never be completely precise. Therefore, judgment must be used in the economic appraisal of investment projects. The value of cost-benefit analysis is that it provides a framework for evaluating both the magnitude of the costs and benefits, and their distribution over time. Such a framework allows the judgments that must be made in assessing the likely yield of an investment to be explicit rather than implicit and possibly vague.

For example, judgments must be made about the real value of the resources to be used in an investment project since their real value may not be fully reflected in their market price because of distortions in the market, such as exchange controls or government control of wages and salaries. Judgments of this type can be incorporated into the appraisal by means of shadow prices, which are intended to reflect the real value of resources to the economy in the light of the social and economic objectives of a country. Shadow prices represent the weight given to different objectives, for example, to future growth as opposed to present consumption.

The World Bank uses the techniques of cost-benefit analysis—and, where appropriate, shadow prices and shadow wage rates—to appraise investment projects. All cost-benefit analysis uses discounted cash flow techniques to compare the discounted present value of both costs and benefits, and to determine whether the benefits accruing from an investment project will be greater than the costs when both are measured in terms of present values. What is needed for such an appraisal is a convenient summary statistic that expresses the relationship between costs, benefits, and their distribution over time. This information can be expressed in three ways, which yield the following investment criteria: the benefit-cost ratio, which is the ratio of the sum of discounted future benefits and the discounted value of costs; the net present value, which is the value of the discounted benefits of a project minus the discounted value of its costs; and the internal or economic rate of return, which is the rate of interest that equates the discounted present value of expected benefits and the present value of costs.

The economic appraisal of investment projects by the World Bank and other development agencies is based on calculations of the net present value of projects and also on calculations of the rate of return. These criteria are never used in isolation to assess the profitability of a project, but they are considered to be one of the essential yardsticks by which alternative investments can be judged (Squire and van der Tak 1975). The question is whether they are applicable to every type of investment, most notably investments in education.

As noted earlier, the idea that education is a form of investment in human capital is one of the most important developments in economics in recent decades, and it has had considerable impact on educational planning in developing countries. At the same time, controversy surrounds the question of whether the techniques of cost-benefit analysis can be used to evaluate investment in education; in particular, doubts have arisen concerning the benefits of investment in education. The purpose of this chapter is to examine the considerable experience with the cost-benefit analysis of education, to explain the techniques that have been used and the objections raised against them, and, finally, to evaluate cost-benefit analysis as a means of assessing investment in education.

The Rate of Return to Investment in Education

In most examples of the cost-benefit analysis of education, calculations have been based on the internal rate of return, rather than on the alternative criteria, benefit-cost ratio or net present value. In the economic literature on investment appraisal, however, the net present value of a project is regarded as a better guide for investment choice than the internal rate of return. The reason is that in some circumstances the two criteria may give conflicting signals, and thus in comparisons of two mutually exclusive projects—for example, the construction of a dam upstream or downstream—the internal rate of return may be misleading. In the case of education, however, the two criteria would seldom provide conflicting advice, and the rate of return has the advantage of being more readily comprehensible.[1] If cost-benefit analysis suggests that the rate of return to expanding university-level education is 10 percent compared with a rate of return of 12 percent for secondary-level education, the relative profitability of the two types of investment can be directly compared. The expansions here are not mutually exclusive projects, as in the example of the dam, even though budget constraints may prevent the government concerned from expanding both levels of education simultaneously. Of course, many uncertainties surround the interpretation of rates of return, as explained below, but these apply with equal force to the interpretation of benefit-cost ratios and net present values, both of which are less intuitively meaningful to the nonspecialist.

Another advantage of using the rate of return is that it is not necessary to build into the cost-benefit analysis any assumption about the rate of interest, or discount, which represents the opportunity cost of capital in the economy, and which must therefore be used to assess the profitability of investment. Benefit-cost ratios and net present value cannot be used without selecting a discount rate at the outset, so that the values of the criteria depend on the choice of discount rate.

Calculation of the rate of return, on the other hand, simply identifies the rate of interest, or discount, which equates the present value of costs and the present value of expected benefits.

If the costs of a project are C_t a year and the project is expected to yield benefits of B_t a year over n years, then the rate of return (r) is the rate of interest at which the present value of costs, $C_t/(1 + r)^t$, is exactly equal in value to the discounted sum of benefits, $B_t/(1 + r)^t$, from year 0 to year n.[2] The calculation of the discounted present value of costs and benefits means that all costs are cumulated forward to year 0, and all benefits are discounted back to the same point in time, as illustrated in figure 3-1. The

Figure 3-1. *A Cost-Benefit Comparison*

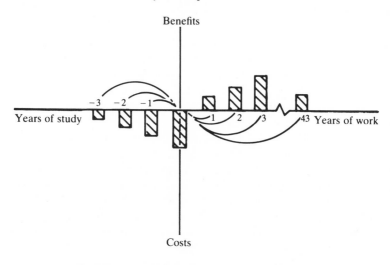

Source: Psacharopoulos (1973), p. 21.

process of discounting takes into account the fact that income expected in the future is worth less than the equivalent amount today, since today's income could be invested at a positive rate of interest and would therefore increase in value. The method of calculation consists of multiplying the value of costs or benefits in each year by a discount factor $1/(1 + r)^t$, using compound interest tables that show the value of $(1 + r)^t$ or of $1/(1 + r)^t$ for any number of years (t).[3]

This type of calculation has been used to measure the rate of return to investment in education in more than fifty countries. There are alternative methods of calculating the rate of return, but this chapter concentrates on the method most widely used, which is to estimate the annual net benefits $(B_t - C_t)$ for each year from 0 to n, to apply alternative discount factors to calculate the present value of net benefits at alternative rates of interest, and to find, either by means of an iterative computer program or simply by trial and error, the rate of interest at which the present value of net benefits is zero. An alternative method is a "short-cut" some studies have used (Psacharopoulos 1981) that gives a less precise result but is useful in cases where the data on costs and benefits are inadequate, since it can provide a rough approximation of the rate of return. Another method that can provide a rough indication of rates of return is based on earnings functions (Mincer 1974), but, like the short-cut method, it makes some simplifying assumptions.[4]

All three methods of calculation yield a single figure—the internal rate of return to investment in education, which is a measure of the profitability of investment from the point of view of individual students or families, or from the point of view of society as a whole. The *private* rate of return measures the relationship between costs and benefits of education for the individual, and the *social* rate of return measures the relationship between all the social costs of education that must be borne by society as a whole, and the benefits that are expected to accrue to society.

Both private and social rates of return are important tools for evaluating investment. Not only is the private rate of return one of the factors that determine individual demand for education, but it also has great bearing on the question of how education should be financed and how the costs and benefits of education should be distributed. The private rate of return is discussed in chapters 5 and 6, which deal with the demand for education and educational finance. The social rate of return—that is, the relationship between the costs and the benefits of education for society at large, rather than for individuals—is the subject of this chapter.

Cost-benefit analysis is an important ingredient of investment appraisal because it helps policymakers decide which of the alternative ways of allocating limited resources will produce the maximum benefits. One of the tools employed in reaching such a decision with respect to investment in education is the social rate of return. This tool is sometimes used to compare investment in education with investment in physical capital, but more frequently it is used to compare the costs and benefits of different types or levels of education. In either case, the rate of return itself must first be compared with the social opportunity cost of capital, which represents the yardstick by which all social investments must be judged. Any project with a rate of return lower than the social opportunity cost of capital should be rejected as unprofitable. The second step is to compare the rate of return of alternative investments. The most profitable project is the one offering the highest rate of return. The crucial step in all cost-benefit comparisons is the identification and measurement of costs and benefits.

The Measurement of Costs

The cost of any investment must be measured by its *opportunity cost*, rather than simply by monetary expenditures. Economic (as opposed to financial) analysis of investment in education thus attempts to estimate the total cost of an investment in terms of alternative opportunities forgone.

This means that it is necessary to identify all the resources, both human and physical, that are used in an education project, and not simply those for which an expenditure item appears in the budget of the Ministry of

Education. Indeed, some of these items may appear in the budgets of other ministries—the Ministry of Transport, for example, may finance school transport, or the Ministry of Agriculture or Development may finance other items of expenditure. If these items are ignored, the costs of investment in education will be underestimated and the investment will appear more profitable than it really is.

Some of the resources used in education do not appear in any budget and thus cannot be captured as an expenditure. For example, a local community may donate land for a school or may provide free food or lodging for teachers. One of the most important resources devoted to education is the time of teachers and pupils or students. The value of teachers' time is measured in wages and salaries, but the value of students' time cannot be measured in the same way. Nevertheless, the fact that students devote their time to education means that it is no longer available for alternative uses such as help in agricultural production. In other words, it has an opportunity cost, though not a financial cost.

The opportunity cost of student time is estimated in terms of the value of the alternative opportunities that are forgone by society; the monetary value of this cost can be derived simply by calculating *earnings forgone*. The wages and salaries that a student must forgo in order to enroll in education rather than find employment represent a cost not only to the individual or to his or her family, but also to society, since they reflect the value of the goods or services that the student could have produced in employment. The earnings forgone by a student in higher education, for example, are usually determined from the average earnings of secondary school leavers who are in employment. In countries with a high level of unemployment, actual earnings may overestimate the opportunity costs of time, since the alternative to education for some students would be unemployment, rather than a wage. Nevertheless, even when unemployment is high, the opportunity cost of students' time is seldom zero. What is relevant is the probability of a student's finding paid employment. Even if unemployment is severe, the probability may be positive, though less than unity. Such probabilities can therefore be used as weights and can be applied to the observed earnings of secondary school leavers to provide an estimate of the opportunity cost of students' time.

Consider, for example, the proportion of school leavers (matriculates) in India who were employed or unemployed one, two, three, and four years after completing their secondary education (table 3-1). The probability of a school leaver being employed in this case was only 35 percent in the first year after completing secondary school, but had risen to nearly 90 percent by the fourth year. This means that if the opportunity cost of a student's time during four years of university education was measured simply in terms of the average wages of workers who had completed

Table 3-1. *Matriculates Unemployed during Each Year after Completing Education, India, 1954*
(percent)

Year after completing education	Percent	
	Unemployed	Employed
1	65	35
2	36	64
3	20	80
4	11	89
5	06	94
6	02	98

Note: It is assumed that within each time interval the number who had found a job grew linearly. The table relates to numbers not having found and having found a *first* job.
Source: Blaug, Layard, and Woodhall (1969), p. 90.

secondary education, it would overestimate the value of students' time. The value of their time in the example was certainly not zero, however. In the first year, the earnings forgone by the average university student were 35 percent of the average wage of school leavers, and by the third year, earnings forgone had risen to 80 percent of the average wages. This example shows one way of using data on unemployment and average wages to estimate the opportunity cost of students' time.

For the individual, forgone earnings often represent the largest proportion of the private costs of education; another proportion consists of fees and expenditure on books. When scholarships, bursaries, or other forms of financial aid reduce the private costs of education, they should be deducted from *private* costs. Expenditure on scholarships or fees should not be included in the estimate of the *social* costs of education, however, since this represents a *transfer payment*, which transfers purchasing power from one group in society to another. Transfer payments do not use up real resources, but simply transfer the power to purchase resources, and therefore do not involve any opportunity costs.

The total cost of the resources that society devotes to education includes the cost of teachers and other staff, books, other goods and services such as heating or lighting, and the value of buildings and equipment. If the land or buildings are already owned by the government, their value cannot be considered an expenditure, but this does not mean that they have no opportunity cost. The opportunity cost of buildings and equipment is usually estimated by means of amortization. If a school building is assumed to have a life of thirty years, say, then the capital cost of the building may be amortized over thirty years to give the annual value of the building. This is sometimes called *imputed rent*. If schools or colleges actually rented buildings, then the annual expenditure on rent would appear as a financial cost in the budget, but even if the buildings are government owned, their use involves an opportunity cost, even though no financial costs are incurred.

Before capital costs can be amortized, certain assumptions must be made about the average life of the building or equipment and the social discount rate that measures the opportunity cost of capital. Some amortization calculations make no allowance for discounting, but simply divide the capital cost of a building by its expected life to arrive at the annual value. This procedure ignores the fact that if capital is tied up in a building or in equipment, it is not available for alternative investments; in other words, here too society incurs an opportunity cost. If, for example, a government has a choice between spending $100,000 to construct a building that will last for twenty years or spending $5,000 a year to rent a building, the total expenditure in either case will be $100,000, but the option of renting the building is economically preferable because the $100,000 could be invested in an alternative project and earn interest for twenty years. If the average rate of return of alternative investment projects is 10 percent, then the calculation of the amortization should use a discount rate of 10 percent to calculate the annual value of the building. Similar arguments apply to the amortization of equipment, for example, in instructional technology projects. (Amortization and the treatment of capital costs are discussed in more detail in chapter 7.)

The important point here is that any estimate of the social costs of education must include a measure of the value of buildings and equipment, and the calculation of amortization will depend on the assumptions made about the opportunity cost of capital and the length of life of the capital. A cost-benefit analysis of educational investment in Kenya carried out for the World Bank (Thias and Carnoy 1972) estimated capital costs for primary schools according to whether the school buildings were permanent, semipermanent, or traditional mud-and-wattle structures. These were assumed to have lives of forty, thirty, and ten years, respectively. Because the amortization calculations in this study were not discounted, however, the opportunity costs would have been slightly underestimated.

Fortunately, the choice of assumptions about the life of capital and the appropriate discount rate for amortization is not usually critical. A study in India that used a range of assumptions about the life of buildings and the appropriate discount rate found the final calculations of rates of return to be relatively insensitive to changes in these assumptions (Blaug, Layard, and Woodhall 1969).

The difference between private and social costs of education (see table 3-2) depends on the extent to which individual students or their families

Table 3-2. *Social and Private Costs of Education*

Social costs	Private costs
Direct	
Teachers' salaries	Fees, minus average value of scholarships
Other current expenditure on goods and services	Books, etc.
Expenditure on books, etc.	
Imputed rent	
Indirect	
Earnings forgone	Earnings forgone

Source: Woodhall (1970), p. 17.

are subsidized by other members of society, either by means of scholarships that cover all or part of tuition fees or earnings forgone, or by means of low or zero fees. Although scholarships constitute a transfer payment and therefore are not included in social costs, the level of government expenditure on scholarships does help to determine the pattern of rates of return to educational investment, through its effect on the disparities between social and private rates of return. (The implications of this disparity will be examined in more detail in chapters 6 and 9.)

A realistic estimate of the social costs of investment in education can only be arrived at if allowance is made for wastage or repetition of classes, a problem that has been studied extensively in developing countries by Unesco and the World Bank (its effects on the internal efficiency of education are examined in chapter 8). If 50 percent of the pupils enrolled in secondary schools were to drop out before completing their schooling, the cost of educating these dropouts would have to be included along with the cost of educating those who successfully complete secondary education. Wastage and repetition increase the social costs of education without correspondingly increasing the benefits. According to World Bank data on wastage rates, a number of developing countries have to provide ten years of primary schooling—instead of a normal or prescribed period of five years—to produce one successful school completer because of dropout and repetition (World Bank 1980a, p. 30). If the average social cost of primary education per pupil-year is $100, then the benefits associated with the successful completion of primary schooling should be compared with a total (undiscounted) cost of $1,000 rather than $500.

It is particularly important to include the social costs of wastage and repetition in these calculations in developing countries where the wastage rates are very high, as illustrated by a study of graduate unemployment in India (Blaug, Layard, and Woodhall 1969). Here it was found that if the costs of wastage are included, the social rate of return to primary education falls from 20 percent to 16 percent. Once the opportunity cost of education to society has been estimated, it can be compared with the expected benefits of the investment by means of the social rate of return.

Identifying and Measuring the Benefits of Education

Education yields direct and indirect benefits both to individuals and to society. The most obvious direct benefit is that educated workers receive higher incomes than those who are less educated. Thus the direct benefit of education for individuals is higher lifetime earnings, and for society it is the higher productivity of educated workers and the additional contributions to national income over their entire working lives. The higher lifetime earnings of educated manpower may be used to measure the direct benefits of education, provided one accepts the critical assumption that the relative earnings of workers reflect their productivity, and therefore that the additional earnings are a proxy measure of the higher output produced by educated workers. Education also yields a set of indirect benefits (known as externalities) that are not immediately captured by the individual and that are extremely difficult to measure empirically (see "Externalities" below).

In the case of private benefits, it is not necessary to make any assumption about the link between education and productivity. If educated workers earn more than uneducated workers, the higher lifetime earnings represent a direct financial benefit to the individual regardless of why employers choose to pay them higher wages. In the case of social benefits, however, the assumption that the higher earnings of educated manpower reflect their higher productivity is a crucial assumption, and its validity is examined below in the light of various objections to cost-benefit analysis. The "screening hypothesis," for example, postulates that employers use educational qualifications simply as a screening device to identify workers with particular abilities, attributes, or attitudes. If this is true, then the higher earnings of educated workers may simply reflect these traits rather than their education. For the moment, however, we leave aside the question of the link between education and productivity to consider how data on earnings are used to calculate age-earnings profiles, which provide a measure of the extra lifetime incomes of educated manpower.

Data on the lifetime pattern of earnings can be obtained in two ways. One is to follow the career of a sample of workers over a period of time to discover how their earnings change. This is known as a longtitudinal, or cohort, study, an example of which is the recent tracer study cited below that attempts to follow a sample of school leavers after graduation. Such longtitudinal studies are in their infancy, however; it would in any case take far too long to collect data on the earnings of cohorts over their entire working lives. The second method, which is the one used in most cost-benefit studies, is to obtain data on the earnings of a sample of workers of different ages at a single point in time. This information can be used to estimate the effect of age or experience on earnings and thus to construct an age-earnings profile. Another advantage of this method is that it avoids the problem of changes over time in the value of money caused by inflation.

The first requirement for an analysis of the effect of education on earnings, then, is that data on the earnings of workers be obtained and classified according to age and educational level. The United States is unusual in that such data have been collected in the national census since 1939, but in most countries it is necessary to conduct a sample survey to obtain data on workers' earnings, age, and educational level or qualifications. From these data it is possible to derive age-earnings profiles for each educational level, as in figure 3-2, which shows age-earnings profiles for a sample of workers in urban India in 1960. Such profiles can then be used to calculate the average lifetime earnings associated with different levels of education.

Figure 3-2. *Age-Earnings Profiles, by Level of Education, Urban India, 1960*

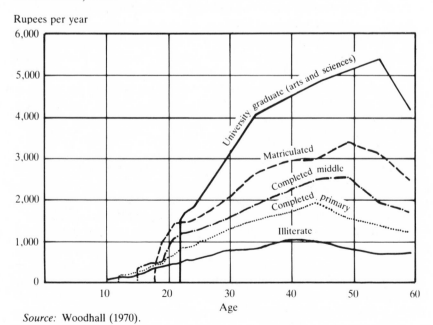

Source: Woodhall (1970).

Age-earnings profiles are available for about fifty countries, and each shows a strong relationship between earnings and education: that is, throughout the world, both in developed and in developing countries, the average lifetime earnings of educated workers are higher than the average earnings of illiterate workers, or of those with lower levels of education. Average earnings tend to rise to a peak in midcareer or later and then stabilize or decline until the age of retirement. The typical characteristics of these age-earnings profiles are:

- Earnings are highly correlated with education; at every age the highly educated earn more than workers with less education, and there is no crossing of profiles.
- Earnings rise with age to a single peak and then flatten or fall until retirement age.
- The profiles are steeper for higher-educated individuals than for those with less education.
- The higher the level of education, the later the age at which earnings reach their peak.

When age-earnings profiles are available for university graduates and workers with secondary education, for example, they can be used to calculate the earnings differentials associated with university education, and the extra lifetime earnings of university graduates compared with secondary school leavers. If the extra earnings of graduates turned out to be entirely due to their education, then their additional lifetime earnings could be used as a measure of the economic benefits of university education, and the total sum of benefits could be calculated simply by adding the earnings differentials of graduates at every age throughout their working lives. Thus in figure 3-3 the extra earnings or benefits (indicated by plus signs), must be compared with the negative area (indicated by minus signs), which represents earnings forgone and the other costs of education. The rate of return can then be calculated as the rate of interest at which the present value of the positive and negative areas in figure 3-3 are equal.[5]

Earnings are determined not only by a worker's educational level, however, but also by age. Furthermore, they reflect other forms of investment in human capital (including on-the-job training), as well as the workers' natural ability, personal characteristics (such as attitudes, motivation, social class, family background, sex, race, place of work—for example, urban or rural), and other variables that influence earning capacity.

The fact that average earnings increase with age indicates that work experience increases workers' productivity. Mincer (1974) has shown that in the United States earnings are more closely correlated with years of work experience than with age. Another study (Simmons 1974) has

Figure 3-3. *A Rate of Return Estimation
for University-Level Education*

Labor earnings (U.S. dollars)

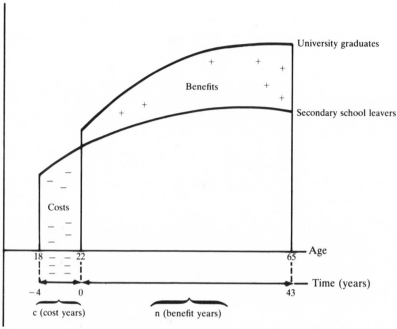

Direct costs (U.S. dollars)

Source: Psacharopoulos (1981), p. 322.

shown that in Tunisia work experience is a more powerful influence on earnings of workers in the shoe industry than are cognitive skills or years of primary schooling; the conclusion here was that work experience is a proxy for informal on-the-job training. Other variables apart from age and work experience also influence earnings, however, and some allowance must be made for these before earnings differentials can be used to measure the benefits of education.

The Influence of Ability and Other Factors on Earning Capacity

The pure effects of education on earnings cannot be identified unless earnings are standardized for other factors, such as natural ability or family background. Regression analysis is one way to measure the effects of other variables such as ability (measured by IQ scores). In the simplest application of regression analysis, observed earnings differentials would be adjusted by a constant factor that represents the proportion of earnings differentials assumed to be the result of other factors, apart from education. This is widely described as the "alpha coefficient" (α), and simply means that if ability plus other factors are assumed to account for one-third of the extra earnings of educated workers while education is said to account for two-thirds, then the extra lifetime earnings of the educated are multiplied by an alpha coefficient of 0.66.

A more satisfactory way of making this adjustment is to estimate an earnings function of the form

$$Y = f(S, IQ, F, \text{Age} \ldots)$$

where Y represents the earnings of an individual, and the amount of schooling (S), natural ability (IQ score or some other index), and family background (F) help to determine earnings, together with age as a proxy for experience.

Regression analysis and earnings functions estimated for workers in the United States and other developed countries suggest that natural ability accounts for slightly less than 20 percent of the additional earnings of educated workers, and that when other factors such as race, sex, and family background are included, education is still the most important single determinant of earnings. A review of the evidence in developed countries shows that the most likely value for the alpha coefficient—in other words, how much of the extra earnings of the educated can be attributed to their education—is about 0.7 or 0.8 (Psacharopoulos 1975).

Less is known, however, about the effects of other factors on earnings in developing countries. Among the studies in this area that have been carried out for the World Bank is a cost-benefit analysis of education in Kenya, which estimated the effect of many variables on earnings, including occupation, family background as measured by father's occupation, parents' literacy, tribal group, urban or rural employment, type of school attended, examination scores, number of years' schooling, and age (Thias and Carnoy 1972). In this case, earnings differentials were standardized for socioeconomic and other variables; when these adjustments are taken into account, the rate of return to investment in education is reduced.

Table 3-3 shows the effect on the social rate of return for urban workers in Kenya when earnings differentials are adjusted for mortality and socioeconomic variables. The socioeconomic variables here appear to be more important in determining the earnings of primary and secondary school leavers than the earnings of those with higher secondary or university education. The social rate of return to higher education is un-

Table 3-3. *Average Social Rates of Return to Schooling, Kenya, 1968*
(percent)

Years of schooling	Urban rates, adjusted for:		
	Age only	Age and mortality	Age, mortality, and socio-economic variables
Primary			
2–4	16	15	11
5–7	38	38	17
2–7	22	21	14
Secondary			
8–9	16	15	19
10–11	34	34	28
8–11	24	24	23
Higher secondary			
12–13	15	15	15
University			
14–17	9	9	9

Note: For each set of adjustments, each rate is recalculated using adjusted costs and benefits.
Source: Thias and Carnoy (1972), p. 92.

affected by these adjustments, whereas the social rate of return to primary education is reduced by standardizing earnings for socioeconomic variables (Thias and Carnoy 1972). This study did not succeed, however, in isolating the effect of natural ability on earnings in Kenya.

A number of recent studies both in developed and in developing countries have used earnings functions to explore the relationship between education (as measured by years of schooling), ability (as measured by IQ), cognitive skills (as measured by test scores), and earnings. One such study for a developing country (Boissière, Knight, and Sabot 1982) suggests that ability, years of schooling, and cognitive skills interact to influence earnings. This conclusion is based on estimated earnings functions for samples of 2,000 workers in Kenya and Tanzania and on their standardized earnings for variables such as reasoning ability and cognitive achievement (as measured by specially administered tests). The results indicate that the indirect effects of reasoning ability and cognitive achievement on earnings differentials exceed the direct effects of years of schooling. Other research, too, suggests that ability and schooling interact, and that their combined effects are greater than their separate effects (Hause 1971).

Under the influence of these relationships, the social rate of return to secondary schooling in Kenya is 12 percent, whereas the unadjusted rate of return is 13 or 16 percent, depending on which sample is used. If it is assumed that the opportunity cost of capital is about 10 percent, then, according to these calculations, secondary education is a profitable investment, and the higher earnings of educated workers do not simply reflect their superior natural ability.

Nevertheless, ability clearly contributes to earning capacity. Some attempts have therefore been made to determine what factors influence the development of ability in young children, and, in particular, whether preschool intervention programs are effective in increasing the ability of children before the age of formal schooling (Grawe 1979). An interesting finding is that the status of the mother, including her educational level, is an important determinant of preschool ability, but the evaluations of the preschool intervention programs have been inconclusive. Even so, they provide further evidence of the complex relationships between ability and education—in this case, there is an intergenerational link between a parent's level of education and the ability of children, which is discussed in more detail below.

Education as a Screening or Filtering Device

Research into the question of how ability, cognitive skills, and years of schooling affect earnings has also shed light on another issue surrounded by fierce controversy: whether the level of education influences productivity and the effect is reflected in earnings, or whether it merely identifies workers with superior ability and personal attributes (such as motivation and attitudes to work, authority, or modernization) and thus is used as a convenient screening device. Critics of the concept of human capital have argued that education may identify productive capacities without necessarily enhancing them. If education does not directly improve workers' skills and productivity, then education may be a profitable private investment, but society derives much less benefit from it. That is to say, education simply confers "credentials" that employers can use to select workers and to determine relative wages and salaries. In the literature on the economics of education this line of reasoning is called "credentialism" or the "screening hypothesis."

At first sight, it looks as though this argument would undermine any cost-benefit analysis of education by denying that earnings measure productivity and that education increases the productive potential of the individual. Evidence on the relationship between education and productivity measured not in terms of earnings but in terms of physical measures of output suggests, however, that credentialism, or the screening hypothesis, is not as damaging as it first appears to be.

In the first place, when a distinction is made between "initial" and "persistent" screening—or what one study describes as "weak" or "strong" versions of the screening hypothesis—it is very hard to find evidence that employers keep paying wages above a worker's productivity *after* the employee has been under their observation for some time (Psacharopoulos 1980b). Initial screening, on the other hand, certainly exists; that is to say, employers may hire someone on the basis of expected productivity, as indicated by the candidate's educational qualifications. But there is nothing wrong with that, since employers need to use selection criteria when hiring workers, and it is both more efficient and equitable for education to be used as a criterion rather than race, religion, caste, or social background. The fact that employers have not been able to find quicker and cheaper methods of identifying productive workers can actually be used as evidence against the screening hypothesis. Furthermore, the fact that age-earnings profiles by level of education diverge rather than converge over time demonstrates that employers continue to pay educated workers more throughout their working lives when they have direct evidence about their productivity and are not forced to rely on education as a screening device.

At the same time, the screening hypothesis has been of some value in emphasizing that education not only imparts vocationally useful knowledge and skills, but also affects attitudes, motivation, and values, all of which help to determine a worker's productivity and employability. After reviewing the evidence of the effects of education on the earning power of individuals, the 1980 *World Development Report* (p. 47) concludes that "Schooling imparts specific knowledge and develops general reasoning skills (its 'cognitive' effects); it also induces changes in beliefs and values, and in attitudes toward work and society ('noncognitive' effects). The relative importance of these effects is much debated, but poorly understood; both are extremely important."

Thus, the belief that education raises the productivity of workers through both cognitive and noncognitive effects is not entirely incompatible with the idea that many employers use education as a convenient screening device. It may be that they do not need the skills directly imparted by education but do value the attitudes and abilities normally associated with education, including the social and communication skills indirectly fostered by education. In developing countries, education has been particularly effective in improving attitudes toward innovations and modernization. In other words, the productivity and screening functions of education are not mutually exclusive, and both bring economic benefits. Nevertheless, it is the link between education and productivity that has attracted most attention and that lies at the heart of the notion of education as investment in human capital.

The Link between Education and Productivity

When earnings are used to measure the benefits of education, two problems immediately arise. First, if labor markets are not competitive, then relative wages are not necessarily a good measure of the relative productivity of educated and less educated workers. Second, earnings cannot be used to measure the benefits of education for workers in the nonwage sector of the economy. The age-earnings profiles used to calculate social rates of return are usually derived from urban labor market surveys, and there is little information on how education affects the earnings of the self-employed or rural incomes. A World Bank study in Kenya that calculated age-income profiles and rates of return to rural and urban education has shown that the incomes of self-employed small landowners in that country do increase with level of education (Thias and Carnoy 1972, p. 59). In rural households, however, the impact of education is greater on nonfarm income than on farm income, and therefore the profitability of education in rural Kenya depends to a large extent on the supply of off-farm jobs.

Since wages may not in some cases be a reliable measure of productivity, it may be preferable to measure the effect of education on physical measures of output, rather than to use wage differentials as a proxy for productivity differences. The most detailed analysis of the link between education and physical measures of productivity has been carried out among farmers in low-income countries. A survey for the World Bank of eighteen studies in these countries explored the relation between education and agricultural efficiency or productivity, measured in terms of crop production (Lockheed, Jamison, and Lau 1980). The review concluded that if a farmer had completed four years of elementary education, his productivity was, on the average, 8.7 percent higher than that of a farmer with no education. The gain in productivity among farmers having four years of schooling rather than none can be seen in figure 3-4 (Jamison and Lau 1982). If allowance is made for the availability of complementary inputs required for improved farming techniques, the effect of education increases when farmers are able to use complementary inputs. In cases where complementary inputs were available, the annual output of a farmer who had completed four years of primary schooling was 13.2 percent higher, on the average, than that of a farmer who had not been to school (table 3-4). The studies also show that, as suggested by Schultz (1964), education is much more likely to have a positive effect in more

Figure 3-4. *Results of Thirty-one Data Sets Relating Schooling to Agricultural Productivity, Unweighted*

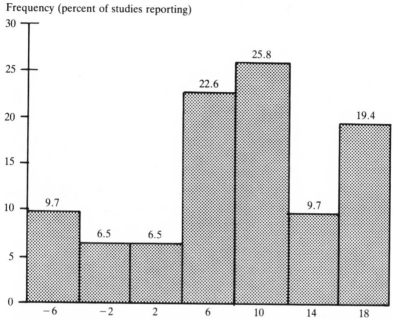

Frequency (percent of studies reporting)

Percentage increase in productivity from 4 years of education

Note: Mean, 8.7 percent; standard deviation, 9.0 percent.
Source: Jamison and Lau (1982), p. 9.

progressive, modernizing agricultural environments, rather than in traditional ones.

Further evidence as to the effect of education on farmers' productivity appears in World Bank studies carried out in the Republic of Korea, Malaysia, and Thailand (Jamison and Lau 1982), and more recently in Nepal and again in Thailand (Jamison and Moock 1984). The first of these studies used estimated production functions for the different types of farms to determine whether education raised the physical output of farmers. The results indicate that the effects of education are "positive, statistically significant and quantitatively important" (Jamison and Lau 1982, p. 8).

In Thailand, data on the prices of farm inputs and outputs were used to test whether education affected allocative efficiency and market efficiency, and whether education affected farmers' choice of production techniques, particularly with respect to the use of chemical fertilizers. According to the analysis, more educated farmers do have higher levels

Table 3-4. *Farmer Education and Farmer Productivity*

Country	Year	Estimated percentage increase in annual farm output due to four years of primary education rather than none
With complementary inputs[a]		
Brazil (Garibaldi)	1970	18.4
Brazil (Resende)	1969	4.0
Brazil (Taquari)	1970	22.1
Brazil (Vicosa)	1969	9.3
Colombia (Chinchina)	1969	−0.8
Colombia (Espinal)	1969	24.4
Kenya	1971–72	6.9
Malaysia	1973	20.4
Nepal (wheat)	1968–69	20.4
Korea, Rep. of	1973	9.1
Average (unweighted)		13.2
Without complementary inputs		
Brazil (Candelaria)	1970	10.8
Brazil (Conceicao de Castelo)	1969	−3.6
Brazil (Guarani)	1970	6.0
Brazil (Paracatu)	1969	−7.2
Colombia (Malaga)	1969	12.4
Colombia (Moniquira)	1969	12.5
Greece	1963	25.9
Average (unweighted)		8.1
No information on availability of complementary inputs		
Average of eight studies (unweighted)		6.3

a. Improved seeds, irrigation, transport to markets, and so on.
Source: World Bank (1980b), p. 48.

of profits (which reflect the higher levels of productivity found in the production function analysis), and higher levels of education and exposure to agricultural extension services do increase the probabilities of using chemical fertilizers.

The education of Thai farmers did not appear to be related to the market prices they obtained for their outputs, however, nor to the prices they paid for their inputs. When Jamison and Lau examined the hypothesis that educated farmers in Thailand would receive higher net prices for their outputs and pay lower net prices for their inputs (since education may enhance a farmer's ability to compare alternatives and may give access to better information), they found no evidence to support this view and concluded that the effect of education on their measure of market efficiency was "rather weak."

Table 3-5. *Internal Rate of Return to Rural Education*
(percent per year)

Farmgate price of rice (U.S. dollars a metric ton)	Age at which benefits are assumed to commence					
	17			25		
	Korea	*Malaysia*	*Thailand*	*Korea*	*Malaysia*	*Thailand*
150	11.0	18.6	17.6	7.1	11.0	10.6
200	13.3	21.8	20.6	8.4	12.5	12.0
250	15.2	24.4	23.2	9.4	13.7	13.2
300	17.0	26.7	25.4	10.2	14.7	14.2

Source: Jamison and Lau (1982), p. 225.

Nevertheless, the fact that farmers' education is positively linked with their physical productivity and choice of technology shows that education provides benefits in rural as well as in urban environments. Moreover, it shows that the idea that education is a profitable investment does not depend on the use of wages as a measure of productivity. As table 3-5 shows, alternative estimates of rates of return to rural education in Korea, Malaysia, and Thailand can be made on the basis of alternative assumptions about the price of rice and the age at which increased productivity actually begins to bring benefits to farmers. Although the figures in table 3-5 are merely rough approximations, they suggest that rural education brings high returns, and this effect does not depend on the assumption that wages reflect productivity. This conclusion is based on the observed relationship between education and productivity, which can be measured in physical terms using the data carefully collected from random samples of farmers.

The fact that such a relationship has been observed does not, of course, explain whether education raises farmer productivity by imparting skills, changing attitudes, or exerting some other influence—or indeed whether education is simply correlated with other attributes that determine farmers' productivity. World Bank research in Nepal and Thailand (Jamison and Moock 1984) is exploring these questions by collecting data on the family background of farmers as well as their education, and by measuring ability, academic achievement, agricultural knowledge, and modernity of attitudes. Analysis suggests that education has an effect independent of family background, and in Nepal this effect appears to be related in part to improvements in farmers' numerical skills. Data on the impact of education on the output of farmers cannot, however, be used to determine whether education directly increases the productivity of urban workers because of differences in the rural and urban labor markets.

Only a few attempts have been made to analyze the effect of education on the physical productivity of urban workers in developing countries. According to a review of such studies for the World Bank (Berry 1980), so far, such studies have proved inconclusive. Also inconclusive are the comparisons that have been made between the benefits of education in urban and rural areas, and in modern and traditional sectors of the economy (which have been used to test the screening or credentialism hypothesis on the grounds that credentialism and screening are likely to be less prevalent in the traditional sector). Comparisons of the benefits of education in the traditional and urban sectors of Colombia and Malaysia, for example, neither support nor refute credentialism. After reviewing all the available evidence, Berry admits that it is impossible to dismiss the idea that credentialism may distort the estimates of the returns to education in some countries, but concludes that the most telling evidence—that is, the substantial returns to education in the informal sector of several countries—would seem to favor the human capital argument that earnings differentials provide a good proxy measure of the benefits of education in most cases.

In other words, although it is likely that education is used partly as a screening device in the urban labor market, this does not invalidate the assumption that education does increase workers' productivity. Furthermore, the evidence that education has a positive effect on farmers' productivity, where the likelihood of screening is much less, lends support to studies of the rates of return to education that assume education offers economic benefits in the form of increased productivity. This does not mean, however, that earnings differentials are a perfect measure of productivity, nor that estimates of rates of return are entirely reliable measures of the profitability of investment in education.

Do Earnings Differentials Measure Productivity?

The idea of perfect competition is central to much of economic theory, even though it is generally agreed that no market is perfectly competitive. For example, economic theory postulates that if labor markets were perfectly competitive, wages would equal the value of workers' marginal product, earnings differentials would be a precise measure of increased productivity, and the extra earnings of educated workers could therefore be used to measure their additional contribution to output. Since labor markets are not perfectly competitive either in developed or in developing countries, however, institutional rigidities may distort the pattern of relative wages. Distortions are particularly evident in developing countries where the public sector is by far the largest employer of educated manpower and many salaries are determined institutionally, rather than being based on market forces.

It has been argued that civil service pay scales in many developing countries reflect historical and administrative influences in their colonial past rather than marginal productivities. In Malaysia, for example, the public sector accounts for 40 percent of total urban employment; furthermore, it employs 90 percent of professional, technical, and related workers, 42 percent of administrative and managerial workers, and 51 percent of clerical workers (Mazumdar 1981). Thus, the majority of those with secondary or higher education are in the public sector. This situation may distort estimates of the rate of return to education if pay scales reflect credentialism and institutional rigidities rather than marginal productivity.

What is at issue, however, is not whether labor markets are perfectly competitive—they clearly are not—but whether market forces have any influence on relative wages. Even in countries where the public sector is the dominant employer, this sector is not completely insulated from market forces (Psacharopoulos 1983). The fear of an upward bias in measurement of the returns to education because of the influence of the public sector should not be exaggerated. Attempts to compare pay scales in the public and private sector in developing countries have produced conflicting results, but even in the public sector, there is evidence in many countries that relative salaries respond gradually to changes in demand and supply. The 1980 *World Development Report* observes, for example, that the relative salaries of teachers and civil servants are much higher in Africa, where educated manpower is scarce, than in Asia, where it is more abundant. The report concludes that, even in the public sector in developing countries, relative wages have gradually but steadily changed in response to increases in the supply of educated labor. Table 3-6 shows how the relative salaries of graduates have declined in several African countries as university education expanded and the scarcity of graduates was reduced. In other words, market forces have influenced relative earnings here, even though a variety of political, social, and historical factors have also played a part in determining salary structures.

The essential condition for rate-of-return analysis is not that labor markets should be perfectly competitive, for this condition is not attained anywhere, but simply that there should be some element of competition, so that relative wages can be interpreted as a signal, though not a perfect measure, of supply and demand. If, however, there are grounds for believing that observed earnings differentials do not provide adequate signals and do not even *reflect* marginal productivity (because of serious distortions in the labor market), then shadow wage rates rather than actual wage rates should be used to estimate the benefits of education.

Table 3-6. *The Changing Relative Earnings of Graduates*
in the Public Sector of Selected African Countries

| Country | Period | Graduate-level salary relative to: | |
		Per capita income	Salary of primary-school graduates
Botswana	1964–74	31.9	10.0
	1974–76	16.7	6.0
Ghana	1967–74	9.5	7.9
	1974–75	11.1	4.8
Kenya	1967–70	25.8	7.0
	1970–74	23.7	6.8
Malawi	1970–71	33.3	11.9
	1975–76	16.7	11.9
Tanzania	1964–65	37.2	8.9
	1970–71	25.8	7.2
Zambia	1970–74	14.1	5.2
	1974–a	13.3	5.1

a. Not specified.
Source: Psacharopoulos (1980a), p. 41.

The Use of Shadow Wage Rates and Prices

It is difficult to estimate shadow wage rates or shadow prices. A certain amount of guesswork is involved since the purpose of shadow prices is to estimate what factors would be paid if their price, or wage, reflected their true economic value. Thus, if distortions in the labor market are so serious that it is estimated that certain groups of workers are paid twice the value of their marginal product, their market wage should be reduced by half to provide a shadow wage rate.

Similarly, if scarcity of foreign exchange means that the official exchange rate in a country underestimates the true value of imported goods and services, then shadow prices should reflect the shadow exchange rate rather than the official exchange rate, which may be kept artificially low through exchange controls. In this case, the shadow exchange rate is an estimate of the exchange rate that would prevail if the price of foreign exchange were allowed to respond to market forces rather than to administrative controls. The World Bank, for example, uses specially calculated conversion factors, which adjust market prices of imported goods or equipment to take account of foreign exchange shortages, in cases where the use of market prices at official exchange rates would distort investment appraisals.[6]

A few attempts have been made to use shadow wage rates and prices to estimate the social rate of return to investment in education in developing countries (Psacharopoulos 1970, Dougherty 1972), but in general, cost-benefit analysis of education has relied on market prices and wages. The various adjustments that have been made to earnings, however—to allow for probabilities of unemployment or the influence of ability or other factors—all resemble attempts to establish shadow wage rates, since they are attempts to improve the reliability of earnings as a measure of the true marginal social product of educated labor.

Externalities

There is one final problem in using earnings differentials to measure the social benefits of education. The earnings of educated individuals do not reflect the external benefits, which affect society as a whole but are not captured by the individual. Such benefits are known as *externalities* or *spillover* benefits, since they spill over to other members of the community. They are often hard to identify and even harder to measure. Nevertheless, cost-benefit analysis should take into account externalities, including the external costs, that may be generated by investment. Such externalities include pollution, congestion, and other undesirable side effects of certain industrial projects, as well as external benefits.

In the case of education, some have succeeded in identifying externalities, but few have been able to quantify them. An early attempt in the United States (Weisbrod 1964) drew attention to the magnitude of externalities, and a recent study (Haveman and Wolfe 1984) concluded that standard rate-of-return estimates may capture only about three-fifths of the full value of education in the United States, including externalities and nonmarket individual benefits. The external benefits of education cited in these studies include crime reduction, social cohesion, technological innovations, and intergenerational benefits (which refer to the benefits parents derive from their own education and transmit to their children). All of these external benefits of education are equally significant in developing countries, where, in addition, education has other important external or spillover effects on fertility and on standards of health and nutrition. An analysis for the World Bank of the relationship between education and fertility (Cochrane 1979) has suggested that a negative correlation between education and fertility is more often observed for the education of females than males. This is particularly important for the cost-benefit analysis of women's education in developing countries, since many women there do not enter the market sector and thus labor earnings do not reflect all the benefits of education for women (Woodhall 1973).

A note of caution concerning externalities and spillovers is sounded in the cost-benefit analysis of education in Kenya carried out for the World Bank (Thias and Carnoy 1972). This study argues that education produces external costs (dis-benefits) as well as benefits. Education may increase national cohesion, for example, but this benefit must be balanced against the possible disintegrating influence of large numbers of unemployed school leavers with frustrated expectations. Similarly, the increased familiarity with modern life of a country's law-abiding majority of citizens is countered by the increased sophistication of its law-breaking minority.

When investment in education is compared with other types of social investment, it should be remembered that all forms of investment may generate externalities. What is important is not whether education as a whole produces spillover benefits, since it clearly does, but whether the externalities for some levels or types of education are greater than others, or whether the externalities of education are more or less significant than they are for other types of investment. Another important question is whether investment in education may help to make other social investment more productive. (We examine the links between education and other sectors in chapter 10, where we show that the effects of education on health, fertility, and agricultural extension services are often very important. All these can be regarded as spillover benefits of education.)

Estimates of the Social Rate of Return to Investment in Education

As noted earlier, estimates of the rate of return to investment in education now exist for more than fifty countries. In the first attempt to carry out a comprehensive international comparison of the rate of return to investment in education (Psacharopoulos 1973), the private and social rates of return to education were compared in thirty-two developed and developing countries. Some of these calculations were based on inadequate data, and some estimated only private rates of return. Nevertheless, the comparison provided a sound basis for analyzing the relationship between costs and benefits of education. It revealed the following general patterns:

- Social returns are consistently lower than the private rate of return.
- Social and private rates of return to primary education tend to be higher than the rate of return to secondary or higher education.
- The rate of return is higher in developing countries than in developed countries.

- The rate of return to investment in education is higher than the average rate of return to physical capital in developing countries, though not necessarily in developed countries.

These patterns were the subject of considerable discussion and controversy, but in some cases they were based on earnings data from the 1950s, which meant that they provided an inadequate basis for assessing the profitability of investment in education in the 1970s and 1980s. The World Bank therefore commissioned an updating of the earlier comparison of rates of return as part of the preparation for the *World Development Report 1980* (Psacharopoulos 1981).

This new comparison—which was based on rate-of-return estimates for forty-four countries (see table 3-7)—reinforced the conclusions of the earlier international comparison in its findings that

- The returns to primary education (whether social or private) are highest among all educational levels.
- The private returns are in excess of social returns, especially at the university level.
- All rates of return to investment in education are well above the 10 percent yardstick commonly used by developing countries to indicate the opportunity cost of capital.
- The returns to education in developing countries are higher than the corresponding returns in more advanced countries.

These conclusions have some important policy implications for the choice of investments in developing countries. First, there is now abundant evidence that education is a profitable social as well as private investment. The fact that the average rate of return in developing countries is considerably higher for primary education than for secondary or higher education (see figure 3-5) suggests that top priority should be given to primary education as a form of investment in human resources. The evidence shows, however, that secondary and higher education are also profitable investments and therefore should be pursued alongside primary education in a program of balanced development of human resources.

Second, the large discrepancy between the private and social returns to investment in higher education has some bearing on financing policy (see chapter 6). Evidence on the rate of return suggests that a shift of part of the cost burden from the state to individuals and their families is not likely to be a disincentive to investing in higher education, given its present high private margin of profitability.

The fact that rates of return are lower in advanced countries suggests that as a country develops, or the capacity of its educational system expands, the returns to educational investment decline. A drastic fall in

Table 3-7. *Returns to Education by Region and Level of Development*
(percent)

| | | Rate of return by educational level | | | | | |
| | Survey | Private | | | Social | | |
Economy	year	Primary	Secondary	Higher	Primary	Secondary	Higher
Developing							
Africa							
Ethiopia	1972	35.0	22.8	27.4	20.3	18.7	9.7
Ghana	1967	24.5	17.0	37.0	18.0	13.0	16.5
Kenya[a]	1971	28.0	33.0	31.0	21.7	19.2	8.8
Malawi	1978	n.a.	n.a.	n.a.	n.a.	15.1	n.a.
Morocco	1970	n.a.	n.a.	n.a.	50.5	10.0	13.0
Nigeria	1966	30.0	14.0	34.0	23.0	12.8	17.0
Sierra Leone	1971	n.a.	n.a.	n.a.	20.0	22.0	9.5
Uganda	1965	n.a.	n.a.	n.a.	66.0	28.6	12.0
Zimbabwe	1960	n.a.	n.a.	n.a.	12.4	n.a.	n.a.
Asia							
India	1965	17.3	18.8	16.2	13.4	15.5	10.3
Indonesia	1977	25.5	15.6	n.a.	n.a.	n.a.	n.a.
Korea, Rep. of	1967	n.a.	n.a.	n.a.	12.0	9.0	5.0
Malaysia	1978	n.a.	32.6	34.5	n.a.	n.a.	n.a.
Philippines	1971	9.0	6.5	9.5	7.0	6.5	8.5
Singapore	1966	n.a.	20.0	25.4	6.6	17.6	14.1
Taiwan	1972	50.0	12.7	15.8	27.0	12.3	17.7
Thailand	1970	56.0	14.5	14.0	30.5	13.0	11.0
Latin America							
Brazil	1970	n.a.	24.7	13.9	n.a.	23.5	13.1
Chile	1959	n.a.	n.a.	n.a.	24.0	16.9	12.2

Country	Year						
Colombia	1973	15.1	15.4	20.7	n.a.	n.a.	n.a.
Mexico	1963	32.0	23.0	29.0	25.0	17.0	23.0
Venezuela	1957	n.a.	18.0	27.0	82.0	17.0	23.0
Intermediate							
Cyprus	1975	15.0	11.2	14.8	n.a.	n.a.	n.a.
Greece	1977	20.0	6.0	5.5	16.5	5.5	4.5
Israel	1958	27.0	6.9	8.0	16.5	6.9	6.6
Iran	1976	n.a.	21.2	18.5	15.2	17.6	13.6
Puerto Rico	1959	n.a.	38.6	41.1	21.9	27.3	21.9
Spain	1971	31.6	10.2	15.5	17.2	8.6	12.8
Turkey	1968	n.a.	24.0	26.0	n.a.	n.a.	8.5
Yugoslavia	1969	7.6	15.3	2.6	9.3	15.4	2.8
Advanced							
Australia	1969	n.a.	14.0	13.9	n.a.	n.a.	n.a.
Belgium	1960	n.a.	21.2	8.7	n.a.	17.1	6.7
Canada	1961	n.a.	16.3	19.7	n.a.	11.7	14.0
Denmark	1964	n.a.	n.a.	10.0	n.a.	n.a.	7.8
France	1970	n.a.	13.8	16.7	n.a.	10.1	10.9
Germany, Fed. Rep.	1964	n.a.	n.a.	4.6	n.a.	n.a.	n.a.
Italy	1969	n.a.	17.3	18.3	n.a.	n.a.	n.a.
Japan	1973	n.a.	5.9	8.1	n.a.	4.6	6.4
Netherlands	1965	n.a.	8.5	10.4	n.a.	5.2	5.5
New Zealand	1966	n.a.	20.0	14.7	n.a.	19.4	13.2
Norway	1966	n.a.	7.4	7.7	n.a.	7.2	7.5
Sweden	1967	n.a.	n.a.	10.3	n.a.	10.5	9.2
United Kingdom[b]	1972	n.a.	11.7	9.6	n.a.	3.6	8.2
United States	1969	n.a.	18.8	15.4	n.a.	10.9	10.9

n.a. Not available.

a. Social rates refer to 1968.

b. Social rates refer to 1966.

Source: Psacharopoulos (1981), pp. 327–28.

Figure 3-5. *The Social Returns to Investment in Education,*
by School Level in Developing Countries

Social rate of return (percent)

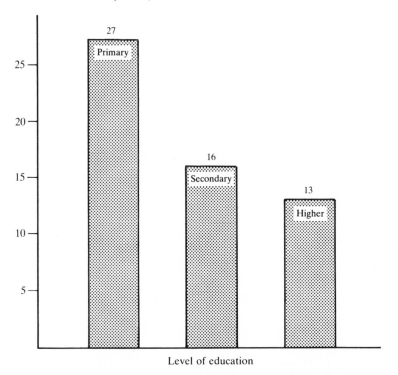

Level of education

Note: "Developing countries" refers to twenty-two African, Asian, and Latin American
countries.
Source: Psacharopoulos (1981), p. 333.

the returns is unlikely, however, as there is some evidence that in countries where a substantial expansion of education has occurred, the rates of
return have declined, but not drastically. A case in point is the United
States, where estimates of the social rate of return have been computed
for every census year since 1939, and estimates of the private rate of
return have been computed annually since 1970 (table 3-8). These time-
series estimates suggest that there has been a gradual decline in the
United States, but not a dramatic one.

Time-series data do not exist for many developing countries. Some
estimates of the private rate of return have been calculated for Colombia,
where a substantial increase in the proportion of the labor force with
secondary and higher education took place between 1964 and 1974; the

Table 3-8. *Time-Series Returns to Education in the United States*
(percent)

	Educational level	
Year	Secondary	Higher
Social rate of return		
1939	18.2	10.7
1949	14.2	10.6
1959	10.1	11.3
1969	10.7	10.9
Private rate of return		
1970	11.3	8.8
1971	12.5	8.0
1972	11.3	7.8
1973	12.0	5.5
1974	14.8	4.8
1975	12.8	5.3
1976	11.0	5.3

Source: Psacharopoulos (1981), p. 335.

evidence suggests that rates of return have fallen relatively slowly in this country (Psacharopoulos 1981). Recent estimates of the social rate of return in Peru and in Burkina Faso (formerly Upper Volta) (see tables 3-9 and 3-10) suggest a sharp decline in the returns to secondary education during the 1970s, but little change in the rate of return to primary and higher education (Psacharopoulos 1982b, c).

The relative stability of rates of return, despite a rapid expansion of education in recent decades, suggests that the demand for educated workers has by and large kept pace with the increased supply generated by educational expansion. This situation may be partly due to technological advances (see figure 3-6), but the influence of technological change on the demand for educated workers is a complex question that still needs further analysis. Although technological change may reduce total demand for labor, it is likely to increase the demand for educated relative to uneducated labor. Thus the rate of return on education—which depends on earnings differentials—may remain high.

The unemployment among educated workers in both developed and developing countries is sometimes used as evidence that education is no longer profitable as an investment, but unemployment does not prove that education is a poor investment. If estimates of benefits are adjusted to take into account probabilities of unemployment, as discussed above, calculations of rates of return already reflect the fact of unemployment. In India, for example, calculations of rates of return that take into account unemployment rates among primary and secondary school leav-

Table 3-9. *Social Rates of Return to Investment*
in Formal Education in Peru
(percent)

Educational level	1972	1974	1980
Primary	46.9	34.3	41.4
Secondary	19.8	9.0	3.3
Higher	16.3	15.0	16.1

Source: Psacharopoulos (1982b), p. 22.

Table 3-10. *Approximate Social Rates of Return to Investment*
in Education in Burkina Faso
(percent)

Educational level	1970	1975	1982
Primary (vs. illiterate)	25.9	27.7	20.1
Secondary (vs. primary)	60.6	30.1	14.9
Higher (vs. secondary)	n.a.	22.0	21.3

n.a. Not available.
Source: Psacharopoulos (1982c).

ers and university graduates show that education is still a profitable investment for the individual even when allowances are made for a significant "waiting period" before employment; and although social rates of return are lower, most are still higher than the opportunity cost of capital in India (Blaug, Layard, and Woodhall 1969). A study for the World Bank challenged the widely held belief that further investment in Indian education would be uneconomic, and concluded: "Despite unprecedented growth, the rates of return (both social and private) remain strong" (Heyneman 1979).

Cost-Benefit Comparisons of Alternative Investments

In addition to the comparisons of rates of return among different levels of education (summarized in table 3-7), there are now a number of examples in which cost-benefit analysis has been applied to different types of education in order to guide investment choices. One of the more important choices educational policymakers in developing countries

Figure 3-6. *A Hypothetical "Race" between Education and Technology That Yields a More or Less Constant Rate of Return to Education over Time*

Rate of return

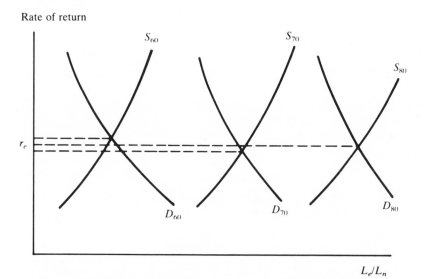

Note: S, supply curve; D, demand curve; L, labor; r, rate of return. Subscripts: *e*, educated; *n*, noneducated; 60, 70, 80 = year.
Source: Psacharopoulos (1981), p. 336.

must make is between a general and a diversified secondary school curriculum.

In the 1970s, Colombia introduced a secondary school system known as INEM (Institutos Nacionales de Educacion Media), which was to provide both an academic and prevocational curriculum that would prepare pupils more effectively for future employment. This experience has recently been the subject of a broad World Bank research project known as the Diversified Secondary Curriculum Study (DiSCuS), which compares both the internal efficiency and the social rate of return to INEM and conventional secondary schools. Here, we look at the comparison between the costs and the benefits of the two types of school (Psacharopoulos and Loxley forthcoming; Psacharopoulos and Zabalza 1984). (The internal efficiency of the diversified curriculum is examined in chapter 8.)

The data for this investigation were collected in a tracer study that consisted of retrospective follow-up of a cohort of about 1,800 graduates of INEM and conventional (control) schools three years after graduation. The purpose of the study was to compare the earnings of graduates from the two types of school to test whether the diversified curriculum yielded higher benefits in the form of increased earnings; differences in earnings were then compared with differences in the cost of the two types of school to give a rough indication of the different social rates of return to investment.

The profitability of the two types of investment could be only roughly estimated since earnings data were available for only a single year (when the graduates were aged 22). The rate of return calculation was therefore based on a short-cut method, which assumes that earnings differentials will remain constant throughout the working lives, as illustrated in figure 3-7. This simplifying assumption allows approximate social rates of return

Figure 3-7. *The Flat–Earnings-Equivalent Assumption for Approximating the Returns to Education*

Earnings

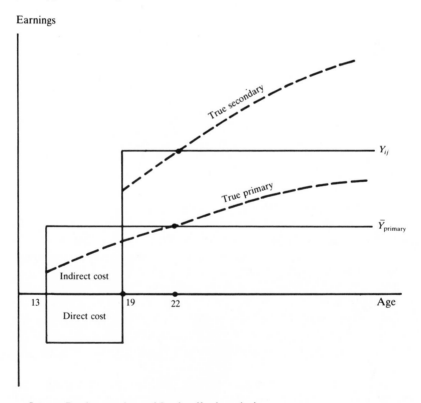

Source: Psacharopoulos and Loxley (forthcoming).

Table 3-11. *Approximate Social Rates of Return to Investment in Secondary Education, by Subject and School Type, Colombia, 1978 Cohort*
(percent)

Subject	INEM		Control	
Academic	7.7		· 9.3	
Agricultural	9.1		7.2	
Commercial	8.4	8.8[a]	9.3	8.3[a]
Social services	7.2		7.2	
Industrial	9.2		9.9	

a. Average social rate of return for nonacademic subjects.
Source: Psacharopoulos and Loxley (forthcoming).

to be calculated for the different types of school and also for different subjects. The results—which are summarized in table 3-11—suggest that there is very little difference in the social rates of return. In chapter 8, we discuss whether the introduction of the diversified curriculum can be justified on grounds of internal efficiency in the schools, as determined by cost-effectiveness analysis. For the moment, however, we must conclude that a cost-benefit comparison between the two types of school does not indicate that INEMs are significantly more or less profitable as an investment than conventional secondary schools.

Another question in developing countries is how best to provide vocational training. There have been fewer attempts to apply cost-benefit analysis to vocational training than to formal schooling, but a number of examples do exist. Social rates of return have been calculated for various types of vocational training in Malaysia, Israel, India, Kenya, Chile, Peru, Brazil, and Colombia, and these have been recently reviewed for the World Bank (Metcalf 1985). Some interesting results emerged from the comparisons. In Israel, for example, in-plant apprenticeship training had a higher rate of return than formal vocational school training, while in Brazil apprenticeship training after primary schooling offered a higher rate of return than junior high school. This result suggests that in some cases vocational training may be a substitute for formal schooling.

The study concludes that such results must be treated with caution, since many problems (such as the interaction between formal schooling and vocational training) arise in cost-benefit comparisons of different forms of training. Nevertheless, this is a promising area for further research, and according to Metcalf the evidence so far suggests that social, corporate, and private returns to vocational training in developing countries are high enough to justify expanding training activity, although no general conclusions can yet be drawn as to the relative merits of training done inside the plant (on or off the job) or outside it.

The Use of Cost-Benefit Analysis of Education

Despite the substantial evidence that now exists for many countries on the rate of return to investment in education, cost-benefit analysis remains a controversial technique, and the objections that have been discussed in this chapter—for example, that earnings do not perfectly measure productivity, that education acts as a screening device, or that cost-benefit analysis does not adequately identify externalities—are frequently cited by skeptics to invalidate estimates of the social rate of return to education. Another problem is that cost-benefit analysis is essentially a tool of marginal analysis and shows the effect of an incremental increase in investment. Furthermore, it may be unreliable as a guide for fundamental, structural changes in the scale or pattern of investment. For all these reasons—and more—cost-benefit analysis cannot be used in isolation as a guide to investment appraisal. Nor should too much trust be placed in numerical estimates of rates of return that in some cases are based on inadequate data or may not take into account wastage, unemployment, or the influence of other factors that help to determine earnings. The rate of return estimates summarized in table 3-7, for example, are unadjusted for the effects of ability or socioeconomic factors on benefits, and also for wastage or unemployment.

Even though rate-of-return estimates cannot claim a high degree of numerical accuracy, however, they can be useful for ranking alternative investments. The fact that cost-benefit calculations consistently show high rates of return to primary education, for example, has led to a reassessment of the economic importance of primary education, both within the World Bank and in other international agencies. It is recognized that high rates of return will not necessarily be maintained in the future as the proportion of workers with primary education increases substantially in developing countries. Nevertheless, there has been a marked shift in Bank lending since 1974, when the second Education Sector Policy Paper argued that an overemphasis on the modern sectors of the economy in many developing countries had caused a misallocation of resources to secondary and higher education at the expense of primary education and the need for education and training in rural areas.

The increase in lending for primary education is certainly not entirely due to cost-benefit analysis, but the evidence from rate-of-return studies in developing countries has helped to focus more attention on the economic benefits of primary education. A review of the evidence on the links between primary schooling and economic development concluded, for example, that primary schooling provides an investment opportunity that ought to have high priority on economic grounds (Colclough 1980).

When in 1962 the World Bank first considered the possibility of lending for education projects, the question of how to assess the profitability of investment in education was raised, and an outline of the criteria for appraising education projects for Bank lending was suggested by a senior staff member. This list included the recommendation that projects be properly planned in human, material, and financial aspects, and that cost estimates and some cost-benefit analysis be included since many of the benefits could not be quantified.

In the twenty years since that was written, considerable progress has been made in the research on cost-benefit analysis of education, some of which has been initiated by the Bank, much of it by individual researchers. An early example of a study commissioned by the World Bank was a discussion of the potential usefulness of a cost-benefit approach to educational planning in developing countries that advocated this type of analysis not as a precise technique, but as a framework for establishing "a scale of priorities which reflect, however crudely, the estimated costs and expected benefits of educational projects" (Blaug 1967, p. 40). This study examined the objections to cost-benefit analysis (particularly the argument that earnings do not perfectly measure productivity), and advocated simple shadow pricing by carrying out sensitivity analysis designed to show what changes in salary scales or relative earnings would be needed to change the ranking of different levels of education. After reviewing all the conceptual problems, the study concluded (p. 25): "Rate of return analysis does not mean slavish adherence to actual earnings and actual costs in total disregard of what earnings and costs mean in a particular country."

In other words, it recommended that evaluations of investment should pay greater attention both to costs and to evidence from earnings and salary patterns, rather than to detailed forecasts of manpower requirements. Although analysis of demand for manpower and of manpower utilization patterns remains a vital part of the Bank's assessment of investment projects (see chapter 4), analysis of costs and benefits and estimates of rates of return now play a greater role than in the past.

The World Bank does not usually make explicit use of calculations of the social rate of return in project appraisal, but rough estimates of rates of return may be used to support the case for investing in education in a particular country. A recent example is a cost-benefit analysis of education in Burkina Faso (Psacharopoulos 1982c), which suggested that the social rate of return to primary education was about 20 percent, to secondary education about 15 percent, and to higher education 21 per-

Table 3-12. *Calculation of Rate of Return on Investment in New School Buildings, El Salvador*

Year	Total cost Option A	Total cost Option B	Cost-saving stream	10 percent Discount factor	10 percent Present value	12 percent Discount factor	12 percent Present value	15 percent Discount factor	15 percent Present value
0	1,485	0	−1,485	1.000	−1,485	1.000	−1,485	1.000	−1,458
1	4,624	0	−4,624	0.909	−4,203	0.893	−4,129	0.869	−4,018
2	4,514	2,458	−2,056	0.826	−1,698	0.797	−1,639	0.756	−1,554
3	4,240	4,036	−204	0.751	−153	0.712	−145	0.658	−134
4–30	3,606	4,921	+1,315(×27)	6.940[a]	+9,126	5.656[a]	+7,438	4.283[a]	+5,632
Total					+1,587		+40		−1,559

Note: Rate of return is approximately 12.5 percent because at this discount rate the net percent value is zero.

a. Compounded 4–30 years.

Source: Sirken (1982).

cent. Since these estimates were based on earnings in the modern private sector and the civil service, they reflect only a small minority of the labor force. The Bank is already planning further research on the benefits of education in the nonwage sector, but in the absence of such research, it would clearly be wrong to recommend the expansion of education in a country such as Burkina Faso simply on the basis of one rate of return calculation. The study shows, however, that cost-benefit analysis is by and large supportive of the Bank's decision to give priority to education as a development strategy in Burkina Faso, and it provides a framework for further analysis of costs and benefits in the future.

Another area in which cost-benefit analysis has proved useful in the Bank's operations is in the evaluation of alternative strategies in a country, for example, in the choice between replacement or improvement of existing school buildings. Such a choice had to be made in El Salvador. Here, the costs of replacing the buildings could be compared with the benefits in the form of reductions in future operating costs (Sirken 1979, 1982). Table 3-12 shows a simple cost-benefit analysis of the two alternatives. Option A, which was to replace the schools, had much higher initial costs, but produced benefits in the form of lower costs from year 4 to year 30 (the assumed life of the schools). When the present value of the cost of replacing the schools (Option A) is compared with the present value of the benefit (the cost-saving shown in table 3-12, column 3), the internal rate of return is found to be 12.5 percent. Thus the replacement of the school buildings was judged to be a profitable investment.

In this example, it was easy to identify the economic benefits of the project; there may, of course, have been other benefits if the higher-quality school buildings affected the efficiency of education in the schools, but this is more difficult to quantify. Even in an apparently simple case where it is easy to identify the benefits, it may not be possible to measure all the costs and benefits precisely; nevertheless, cost-benefit analysis provides a useful framework for comparing potential costs and benefits.

Cost-benefit analysis was also used to compare the rate of return to different types of higher education. A World Bank study of the costs and benefits of university education in developing countries (Psacharopoulos 1982a) suggested that vocational subjects such as agriculture may have a lower rate of return than general university education, partly because the very high costs of agricultural courses are not matched by higher earnings of agricultural graduates. This rather surprising finding has two important implications. First, the rate of return of high-cost subjects such as agriculture or engineering would be even higher if the costs could be reduced. (We consider the scope for cost reductions in education in chapter 7 and the use of cost-effectiveness analysis to improve efficiency in chapter 8.)

Second, the high rates of return to general university education serve as a reminder that although in the past the emphasis in many developing countries and in World Bank lending has been on technical and vocational higher education, general education also has its place in the development of a balanced higher educational system and remains a profitable investment in many countries.

Our general conclusion is that cost-benefit analysis cannot be used in isolation to judge the profitability of investment in education, but it provides one essential ingredient in the assessment and choice of investment projects.

Notes to Chapter 3

1. For a more detailed discussion of the merits of net present value and internal rate of return estimates, see Squire and van der Tak (1975), pp. 39–43, and Blaug, Layard, and Woodhall (1969), pp. 235–36.

2. The internal rate of return is the rate of return at which the present value of costs

$$\sum_{t=0}^{t=n} \frac{C_t}{(1+r)^t}$$

equals the present value of expected benefits

$$\sum_{t=0}^{t=n} \frac{B_t}{(1+r)^t}$$

or, alternatively, the rate of interest (r) at which the difference between discounted benefits and costs is zero, that is,

$$\sum_{t=0}^{t=n} \frac{B_t - C_t}{(1+r)^t} = 0.$$

3. Compound interest and discount factors, for $n = 1$ to $n = 8$, at 10 percent interest are:

Year	Amount to which $1 invested will grow at end of each year	Amount that $1 promised at end of each year is worth today
1	1.100	0.909
2	1.210	0.826
4	1.331	0.751
4	1.464	0.683
5	1.611	0.621
6	1.772	0.564
7	1.949	0.513
8	2.144	0.466

A sum of money (A), invested at a positive rate of compound interest (r) for n years, will grow to $A(1 + r)^n$ by the end of the period. Thus \$1 invested for four years at 10 percent grows to $\$1(1 + 0.10)^4 = 1.464$. The present value of a sum of money (A), expected at the end of n years, when the discount rate is r, is $A/(1 + r)^n$. Thus, \$1 expected at the end of four years, at a discount rate of 10 percent, is now worth $\$1/(1 + 0.10)^4 = 0.683$.

4. For a discussion of the short-cut and earnings function method of calculating rates of return, see Psacharopoulos (1981).

5. In other words, the rate of return is the rate of interest at which discounted benefits to age 22 = cumulative costs at age 22

$$\sum_{t=0}^{t=43} \frac{B_t}{(1 + r)^t} = \sum_{t=-4}^{t=0} \frac{C_t}{(1 + r)^t}$$

or the area indicated by plus signs = the area indicated by minus signs in figure 3-3.

6. For further discussion of shadow prices see Squire and Van der Tak (1975).

References

Berry, Albert. 1980. Education, Income Productivity and Urban Poverty. In *Education and Income*, ed. Timothy King. World Bank Staff Working Paper no. 402. Washington, D.C.

Blaug, Mark. 1967. A Cost-Benefit Approach to Educational Planning in Developing Countries. Economics Department Report no. EC-157. Washington D.C.: World Bank.

Blaug, M., P. R. G. Layard, and M. Woodhall. 1969. *The Causes of Graduate Unemployment in India*. London: Allen Lane, Penguin Books.

Boissière, M., J. B. Knight, and R. H. Sabot. 1982. Earnings, Schooling, Ability and Cognitive Skills. Washington, D.C.: World Bank, Development Research Department. Processed.

Cochrane, Susan H. 1979. *Fertility and Education: What Do We Really Know?* Baltimore, Md.: Johns Hopkins University Press.

Colclough, Christopher. 1980. *Primary Schooling and Economic Development: A Review of the Evidence*. World Bank Staff Working Paper no. 399. Washington, D.C.

Dougherty, C. R. S. 1972. The Optimal Allocation of Investment in Education. In *Studies in Development Planning*, ed. H. B. Chenery. Cambridge, Mass.: Harvard University Press.

Grawe, Roger. 1979. *Ability in Pre-Schoolers, Earnings and Home Environment*. World Bank Staff Working Paper no. 322. Washington, D.C.

Hause, J. C. 1971. Ability and Schooling as Determinants of Earnings, or If You're So Smart Why Aren't You Rich? *American Economic Review* 61 (May):289–98.

Haveman, R., and B. Wolfe. 1984. Education and Economic Well-Being: The Role of Non-Market Effects. *Journal of Human Resources* 19(3):377–407.

Heyneman, Stephen P. 1979. *Investment in Indian Education: Uneconomic?* World Bank Staff Working Paper no. 327. Washington, D.C.

Jamison, Dean T., and Lawrence J. Lau. 1982. *Farmer Education and Farm Efficiency*. Baltimore, Md.: Johns Hopkins University Press.

Jamison, D. T., and P. R. Moock. 1984. Farmer Education and Farm Efficiency in Nepal: The Role of Schooling, Extension Services, and Cognitive Skills. *World Development* 12, no. 1 (January):67–86.

Lockheed, Marlaine E., Dean T. Jamison, and Laurence J. Lau. 1980. Farmer Education and Farm Efficiency: A Survey. In *Education and Income*, ed. Timothy King. World Bank Staff Working Paper no 402. Washington, D.C.

Mazumdar, Dipak. 1981. *The Urban Labor Market and Income Distribution: A Study of Malaysia*. New York: Oxford University Press.

Metcalf, David. 1985. *The Economics of Vocational Training: Past Evidence and Future Evaluations*. World Bank Staff Working Paper no. 713. Washington, D.C.

Mincer, Jacob. 1974. *Schooling, Experience and Earnings*. New York: National Bureau of Economic Research.

Psacharopoulos, George. 1970. Estimating Shadow Rates of Return to Investment in Education. *Journal of Human Resources* 5, no. 1 (Winter):34–50.

———. 1973. *Returns to Education: An International Comparison*. Amsterdam and New York: Elsevier.

———. 1975. *Earnings and Education in OECD Countries*. Paris: Organisation for Economic Co-operation and Development.

———. 1980a. *Higher Education in Developing Countries: A Cost-Benefit Analysis*. World Bank Staff Working Paper no. 440. Washington, D.C.

———. 1980b. On the Weak versus the Strong Version of the Screening Hypothesis. *Economics Letters* 4:181–85.

———. 1981. Returns to Education: An Updated International Comparison. *Comparative Education* 17(3):321–41.

———. 1982a. The Economics of Higher Education in Developing Countries. *Comparative Education Review* 26, no. 2 (June):139–59.

———. 1982b. Peru: Assessing Priorities for Investment in Education and Training. Washington, D.C.: World Bank, Education Department.

———. 1982c. Upper Volta: Is It Worth Spending on Education in a High-Cost Country? Washington, D.C.: World Bank, Education Department.

———. 1983. Education and Private versus Public Sector Pay. *Labor and Society* 8, no. 2 (April–June):123–34.

Psacharopoulos, George, and William Loxley. Forthcoming. *Diversified Secondary Education and Development: Evidence from Colombia and Tanzania*. Baltimore, Md.: Johns Hopkins University Press.

Psacharopoulos George, and Antonio Zabalza. 1984. *The Effect of Diversified Schools on Employment Status and Earnings in Columbia*. World Bank Staff Working Paper no. 653. Washington, D.C.

Schultz, Theodore W. 1964. *Transforming Traditional Agriculture*. New Haven, Conn.: Yale University Press.

Simmons, John. 1974. *The Determinants of Earnings: Towards an Improved Model*. World Bank Staff Working Paper no. 173. Washington, D.C.

Sirken, Irving, ed. 1979. Education Programs and Projects: Analytical Techniques, Case Studies and Exercises. Washington, D.C.: World Bank, Economic Development Institute. Processed.

————, ed. 1982. Education Programs and Projects: Analytical Techniques, Solutions. Washington, D.C.: World Bank, Economic Development Institute. Processed.

Squire, Lyn, and Herman G. van der Tak. 1975. *Economic Analysis of Projects*. Baltimore, Md.: Johns Hopkins University Press.

Thias, H. H., and M. Carnoy. 1972. *Cost-Benefit Analysis in Education: A Case Study on Kenya*. Baltimore, Md.: Johns Hopkins University Press.

Weisbrod, B. 1964. *External Benefits of Public Education: An Economic Analysis*. Princeton, N.J.: Princeton University, Industrial Relations Section.

Woodhall, Maureen. 1970. *Cost-Benefit Analysis in Educational Planning*. Paris: Unesco/International Institute of Educational Planning.

————. 1973. The Economic Returns to Investment in Women's Education. *Higher Education* 2(3):275–99.

World Bank. 1980a. *Education*. Sector Policy Paper. Washington, D.C.

————. 1980b. *World Development Report 1980*. New York: Oxford University Press.

4

Analysis of Demand for Manpower

The basic belief underlying attempts at manpower planning is a simple one: skilled manpower is one of the most crucial inputs of a modern economy. Thus, to foster economic growth (and to avoid critical shortages or surpluses of manpower), planners have sought to identify future requirements for skilled manpower and to design the education systems so as to produce a labor force with the necessary skills and technical or professional knowledge.

The idea that a country's future manpower structure can be predicted and the forecasts used as a basis for planning the scale of education is intuitively appealing, since it appears to offer unambiguous guidance to the policymaker on how to plan educational investment. When the notion that "it is possible to ascertain the optimum amount of education for achieving specified growth targets" was put forward by Parnes (1962, p. 7), he regarded it as "novel," but it has since then exerted a powerful influence over educational and economic planning in many developing countries. According to a survey by Unesco (1968), sixty of seventy-three countries that had drawn up educational plans had attempted to base these plans on forecasts of future manpower requirements. During the 1970s even more developing countries were doing the same.

In some cases, forecasts indicate the general demand for educated manpower by the number of university graduates and primary school and secondary school leavers who will be needed. In other cases, forecasts may reflect the demand for scientific and technical manpower or for highly qualified manpower (classified in terms of occupational categories), or for specific occupations, such as teachers or doctors. The use of such forecasts in educational planning has met with some criticism, however, particularly from those who question not only the reliability of past forecasts, but also the whole notion that manpower requirements might be determined (see, for example, Hollister 1964; Ahamad and Blaug 1973; Psacharopoulos 1984b). The heart of the controversy is not whether economic planning should take into account employment trends and estimates of the demand for and supply of skilled manpower, but whether it is possible or desirable to attempt long-term forecasts, and whether indeed the notion of manpower requirements or needs is a valid one.

Advocates of manpower forecasting argue that because it takes many years to turn out highly skilled or trained manpower, and because shortages of manpower constitute a serious bottleneck that will frustrate development plans, long-term forecasts of manpower requirements are needed to ensure that the education system produces the right combination of skills. Critics argue that the labor market is sufficiently flexible to make the idea of fixed manpower requirements meaningless and that forecasting techniques are unreliable. What is needed, according to them, is analysis of manpower trends, including both existing patterns of manpower utilization and the implications of alternative economic targets, rather than forecasts that may prove inaccurate.

This controversy has mounted in recent years. Much of it has centered on whether investment in education should be guided by manpower forecasting or cost-benefit analysis. Despite pleas for more than fifteen years that the two approaches should be combined, they still tend to be regarded as competing techniques, as two recent reviews of the literature on educational investment illustrate (Dougherty 1983; Hollister 1983). The overall conclusion of the study in which these reviews appear (Psacharopoulos et al. 1983), however, was that neither technique could be completely relied on and that in many situations neither would be applicable. What is needed, it argued, is not a technique or a formal mechanistic model, but an approach to manpower analysis that provides constant feedback and monitoring information, including that derived from rate-of-return estimates and analysis of wage and salary data, as well as analysis of labor market trends. The controversy itself has taken up too much time, the study added, and has diverted attention and resources away from wider aspects of manpower analysis that could provide more useful information and guidance for policymakers.

This controversy can be summarized briefly. Advocates of a manpower requirements approach do not believe relative prices can be a reliable guide to future investment decisions and prefer to use projections of numbers employed, whereas advocates of a rate-of-return approach mistrust purely quantitative forecasts of manpower and use relative wages and salaries to provide signals of demand and supply. Hollister (1983) argues that both dimensions should be taken into account in trying to understand how the market for manpower is likely to evolve in the future. This conclusion echoes the advice given in an earlier comparative study (Blaug 1967), which argued that project evaluation should be based on estimates of manpower supply and demand combined with rates of return.

Nevertheless, the fact that both Hollister (1983) and Dougherty (1983) considered it necessary to treat manpower forecasting and cost-benefit analysis as competing techniques is evidence that this advice has not been followed. Analysis of manpower issues has too often been narrowly restricted to mechanical attempts at manpower forecasting, or to rather generally stated needs for skilled and semiskilled workers, and has made no attempt to examine systematically the actual placement and wages or salaries of graduates, or to compare the costs and effectiveness of alternative combinations of skills or different ways of developing skills (for example, by alternative combinations of formal education and vocational training). Such concerns are well documented in a recent review of manpower planning in Asian countries (ILO 1979).

This chapter therefore focuses on the wider aspects of manpower analysis and their implications for investment projects. The assumptions and techniques of manpower forecasting will be briefly discussed, however, since they have held sway for so many years in the World Bank and in other international agencies, including Unesco, the Organisation for Economic Co-operation and Development (OECD), and the ILO, as well as in individual developing countries.

The Concept of Manpower Requirements

Most advocates of manpower forecasting prefer the term *requirements* or *needs* to the economic term *demand*. The former, they contend, reflects the basic concept behind their approach, whereas the latter denotes the quantity of goods or services that consumers or producers are willing to buy at a particular price and the term therefore emphasizes the close relationship between quantity and price. Requirements or needs, on the other hand, suggest that there is a minimum quantity of skilled labor that is necessary for the achievement of a target level of output or production. Parnes (1962) therefore suggested that the concept of manpower requirements is more a technological than an economic one. The idea that a certain level of skilled manpower is necessary to achieve a particular level of output or economic target rests, basically, on two other assumptions: a fixed relationship between the input of skilled manpower in different occupational categories in an industry, sector, or the economy as a whole and its level of output; and a fixed relationship between the educational qualifications of workers and their productivity.

If labor markets could adjust quickly to changes in demand for different types of workers and if the educational system could adjust quickly by producing workers with the necessary knowledge or skills, then there would be no need, according to one argument, to make long-term forecasts of manpower demand or supply, since both demand and supply would be brought into balance by changes in the pattern of wages and salaries. In the event of a temporary shortage of particular categories of workers, wages would increase and thus would induce more workers to enter the occupation experiencing the shortage. This line of reasoning assumes that there is competition in labor markets; that employers and workers—and indeed students—respond to market forces and have access to information on shortages or surpluses; that workers can acquire the necessary skills quickly; and that labor is mobile. The alternative view is that planners must make long-term forecasts of manpower requirements because labor markets do not fully satisfy these conditions, and there may be considerable time lags between the first signs of a shortage and any increase in wages or salaries. Indeed, if salaries are administratively determined, relative wages may not adequately reflect shortages at all. Furthermore, since it takes many years to produce a skilled teacher or engineer, a shortage of skilled manpower will not lead immediately to an increase in the supply of workers with the necessary skills.

Some economists believe that in such circumstances cobweb cycles can arise in which imbalances between demand and supply may lead to extreme fluctuations in wage rates, which in turn may lead consumers or producers to overreact to shortages or surpluses, as illustrated in figure 4-1. Others, however, believe that both employers and students are able

Figure 4-1. *The Effect of Cobweb Cycles on Demand and Supply*

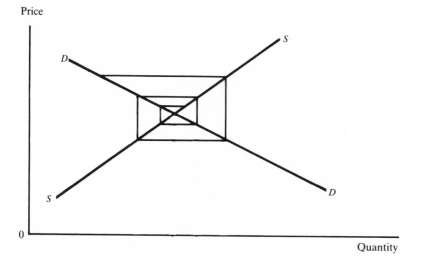

to react more quickly than this model implies. A study of the market for college-trained manpower in the United States (Freeman 1971), for example, concludes that students do respond to market signals, and though there are time lags, they are much less rigid than might be expected. This controversy has generated considerable literature in the economics of education, particularly in the United States, but much of it is not directly relevant to developing countries, and we make no attempt to review it in depth.

The main argument in favor of long-term forecasts of manpower requirements in developing countries is that such time lags in response to shortages or surpluses are wasteful and inefficient and must be avoided by ensuring that the supply of workers with particular skills or educational qualifications exactly matches the demand for their services. According to critics of manpower forecasting, on the other hand, even in developing countries labor markets are sufficiently flexible to adjust to changing patterns of supply and demand in the long run, and it is possible for employers to achieve a desired level of output with a wide variety of input combinations, depending both on production techniques and on relative prices.

The debate about the validity of manpower forecasting thus seems to revolve about the labor market and the degree to which it is flexible, particularly in developing countries. If, as forecasts assume, there are fixed or stable relationships between the input of different categories of manpower and output, and if there is a close link between educational qualifications and occupational structure, then future demand for manpower can be predicted on the basis of existing occupational and educational distributions and assumed targets for productivity or economic growth. If, on the other hand, there is a high degree of substitutability between different inputs, so that the same level of output can be achieved with alternative combinations of labor and capital, skilled and unskilled labor, and if employers can choose between hiring highly educated workers and providing on-the-job training for less educated workers, then it is both more difficult and less necessary to attempt long-term forecasts. It is more difficult because the pattern of employment will depend on a wide variety of factors, including relative prices, rather than on fixed relationships between input and output. It is less necessary because employers will be free to respond to shortages or surpluses of manpower by changing the combination of inputs, for example, by changing the balance of education and training of their work force.

Blaug (1967) believes the controversy reflects "two views of the state of the world" (see table 4-1). The real world obviously lies somewhere between these extremes. What is at issue is whether the labor market and educational systems of developing countries are more characterized by

Table 4-1. *Two Views of Education and the Labor Market*

Extreme version of the manpower-planning approach	*Extreme version of the anti–manpower-planning approach*
Students acquire more education for noneconomic reasons only.	Students acquire more education for economic reasons only.
Students choose major subjects in ignorance of, or with no regard to, career prospects.	Students are well informed and attentive to career prospects.
All education is specialized and specialization starts at age of entry.	All education is general and there is no specialization at any age.
All input coefficients in schools are fixed: complete indivisibility and specificity of teachers, plant, and equipment.	All input coefficients in schools are variable: complete divisibility and nonspecificity of teachers, plant, and equipment.
The demand curves for separate skills shift discretely.	The demand curves for separate skills shift continuously.
Zero elasticities of substitution between skilled labor.	Infinite elasticities of substitution between skilled labor.
Zero elasticities of demand for all separate skills.	Infinite elasticities of demand for all separate skills.

Source: Blaug (1967), p. 32.

fixed relationships and rigidities in both educational and employment practices, or by flexibility and substitutability between different skills and different ways of acquiring skills.

During the 1960s many countries produced forecasts of manpower requirements. Because they assumed there was little flexibility and the right number of engineers or technicians, university graduates or secondary school leavers had to be educated to achieve production targets or expand economic growth, these early manpower forecasts were preoccupied with avoiding shortages. As education expanded rapidly and economic growth and employment failed to match the optimistic targets set in the 1960s and early 1970s, the concern of many planners shifted to the question of how to avoid surpluses and how to reduce unemployment among the educated. Nonetheless, the primary purpose of the manpower forecasts was to provide the planners of educational investment with precise numerical targets that were based on estimates of the numbers of educated workers needed to produce the desired level of output in particular industries or sectors or in the economy as a whole.

Thus, most requests for educational loans from the World Bank and most project appraisals are justified on the grounds of manpower requirements. Whether or not these include detailed forecasts, they rest on the assumption that it is possible to predict manpower requirements. In other words, they assume fixed relationships between education, occupation, and output. The remainder of this chapter looks at the considerable body of research that deals with the techniques and underlying assumptions of manpower forecasting.

Methods of Manpower Forecasting

There is no single method of forecasting manpower requirements. The earliest attempts—such as the Ashby Report in Nigeria (Ashby 1960)—were based on simple rules of thumb, such as the assumption that high-level manpower should grow twice as fast as the target rate of economic growth and that intermediate manpower (workers who had completed secondary schooling and had at least one year of postsecondary education) should grow three times as fast. These rules of thumb were not based on evidence or analysis, but on judgment. More recently, countries have begun to rely on four other techniques: employers' estimates of future manpower requirements, international comparisons, manpower-population ratios, and extrapolation of fixed input-output ratios.

An example of the first technique is an exercise carried out in Nigeria in 1963 and 1965, during which employers were asked to forecast how many workers of different categories they would employ in 1966, 1968, and 1970 (Hinchliffe 1973). The striking result of this experience of employers' estimates is the great variation that occurred over a two-year period. These estimates proved unreliable for many reasons, among them a lack of guidance concerning the assumptions employers should make about price and wage levels, demand for their products, or their own share of the market; thus, they may have made mutually inconsistent assumptions. This illustrates one of the problems of using employers' estimates as a basis for manpower forecasting—namely, that individual employers find it as difficult as governments to make accurate forecasts.

It also shows, however, that information derived from surveys of employers can throw light on the assumptions and techniques of forecasting. One of the methods used by the National Manpower Board in Nigeria to produce forecasts of different categories of manpower was to assume a skill-mix consisting of constant ratios between highly qualified, intermediate qualified, skilled, and unskilled workers. As table 4-2 shows, however, the Nigerian survey of employers indicated considerable variation in the ratios of qualified and skilled workers to unskilled workers. Some of this variation may be due to differences in the classification of workers or to nonresponse (for example, some large firms did not respond to the 1965 survey). It does, however, cast doubt on the assumption that skill-mix is constant over time.

Table 4-2. *Skill-Mix Ratios in Four Sectors, Nigeria, 1963–65*

Sector	Year	Senior	Intermediate	Skilled	Residual
Manufacturing	1963	1	2.2	13.0	15.3
	1964	1	2.3	13.5	15.1
	1965	1	1.2	8.8	7.6
Mining	1963	1	1.4	17.0	22.0
	1964	1	1.4	17.5	22.4
	1965	1	0.97	47.6	11.5
Construction	1963	1	1.7	15.9	17.8
	1964	1	1.7	16.4	17.1
	1965	1	3.7	16.0	25.6
Electricity	1963	1	1.2	8.3	14.7
	1964	1	1.2	7.5	13.3
	1965	1	3.3	7.8	8.4

Source: Ahamad and Blaug (1973), p. 149.

Admittedly, this example dates back to a time when manpower fore-casting was a new activity in Nigeria. Despite the problems described above, however, some developing countries still use employers' esti-mates to forecast manpower needs. This is the case particularly in sectors that are dominated by a few large employers whose opinions can easily be surveyed and who exert a powerful influence over manpower demand in the sector as a whole. One such sector in many countries is energy. In Indonesia, for example, the Ministry of Mining and Energy asked for World Bank assistance to evaluate the present and prospective employ-ment situation in the energy sector, where shortages of skilled manpower could interfere with the economic growth of the country (the sector's share of GDP is expected to rise from 1.3 percent in 1980 to 10.4 percent by 2000). A survey requested information on employers' actual man-power situation in 1981 and on their perceived manpower needs in 1985 and 1990 (Psacharopoulos 1984a).

One of the purposes of this exercise was to discover whether a bot-tleneck is likely to occur in the supply of skilled manpower in Indonesia and possibly limit the development of the energy sector. The quantitative estimates of manpower needs, in relation to estimates of total supply, suggest that such a bottleneck will not necessarily develop, although in certain disciplines (for example, geology, electrical engineering, and accountancy), the estimated manpower needs of the energy sector would absorb a significant proportion of the total supply of graduates in these fields. This exercise showed, however, that quantitative estimates of needs by themselves are not enough to prove or disprove that a shortage will occur. The final balance between demand and supply will depend not on mere numbers, but on whether the state energy firms in Indonesia can attract sufficient qualified workers. This will depend on the relative salaries offered in the public and private sector, on the possibility of creating skills through on-the-job training rather than formal education, and on manpower demand in other sectors of the economy.

A comparison of public and private sector pay scales suggests that the alleged shortage of skilled manpower in public sector energy firms in Indonesia reflects the lower salaries paid in the public sector and the lack of mobility between the private and public sectors. Even if higher education was expanded to increase the supply of graduates in accountancy or geology, there is no guarantee that they would be employed in the energy sector or that they would go to public rather than private employers. This problem, of course, goes beyond employers' estimates of manpower needs; it undermines any attempts at manpower forecasting.

In many occupations in developing countries, however, the public sector is virtually the only employer. The most obvious example is teachers and doctors, where governments control both supply and demand. Yet even in this case considerable problems arise in forecasting future numbers accurately, as we show below.

The second method of forecasting that is commonly used is based on international comparisons. These may consist of simple comparisons of the manpower and educational structures of countries at different stages of development. The assumption here is that to achieve faster economic growth less developed countries should copy the structures found in countries with a higher national income. Alternatively, detailed analysis of input-output relationships in different countries can be used to extrapolate trends, for example, in labor productivity or in occupational and educational distributions.

In either case, the assumption is that different countries have certain manpower growth paths in common and that to achieve a higher level of output a developing country must increase the number of qualified workers employed until the proportion resembles that of a more advanced country. Several attempts have been made to use international comparisons, one of the first being a study by the Netherlands Economic Institute (1966), which was based on data from twenty-three countries. This study concluded that a 1 percent increase in national income was associated with a 1.038 percent increase in high-level manpower and a 0.655 percent increase in manpower with secondary-level schooling or intermediate qualifications. This conclusion was used to forecast manpower requirements in Africa, Asia, and Latin America, where it was assumed that these economic targets could be achieved by increasing stocks of highly qualified manpower at roughly the same rate as the target rate of economic growth and intermediate-level manpower at roughly two-thirds of the target rate of income growth. This gave rise to another rule of thumb that had considerable influence on educational planning, even though the original analysis did not prove either that correlation implied causation or that the educational structures in the twenty-three countries in 1957 were optimal.

More detailed studies of international comparisons by Layard and Saigal (1966), Horowitz, Zymelman, and Herrnstadt (1966), and OECD (1970) investigated the relationship between manpower patterns and level of economic development with a view to forecasting occupational or educational distributions. The most comprehensive of these studies (OECD 1970) examined data from fifty-three countries (the data were classified according to eight economic sectors, ten occupational categories, and four levels of education). This study concluded that in many cases differences in occupational or educational patterns between countries at similar levels of economic or technological development reflect the influence of supply rather than demand, and that substitutions are possible between different types of labor (classified by occupation or education).

Nevertheless, manpower forecasting is still often based on international comparisons. In Korea, for example, the Ministry of Science and Technology has based its forecasts of demand for scientific and technical manpower on the occupational structure of industries in Japan, the United States, and the Federal Republic of Germany. Within the World Bank, too, such comparisons have been suggested as a basis for manpower forecasting. In one analysis of the occupational distribution in 26 countries (Zymelman 1980a, b), data for 120 occupations in 58 industries were used to construct regression equations, as a basis for manpower forecasting. These equations did not take into account differences in relative prices or other economic factors in different countries, however, and to use them to project future occupational structures would imply a fixed occupational pattern over time and across countries. On the other hand, Zymelman (1980b, p. 15) argued that to apply the distribution of education within a given occupation in one country to determine future educational requirements in another would be to apply a "petrified chronicle of the past."

The jump from what is observed in a number of countries to what ought to be achieved in another is a fundamental one that advocates of manpower forecasting are willing to make, but its critics are not. Later sections of this chapter consider whether occupational/education coefficients are in fact stable over time or whether opportunities for flexibility and substitutability exist. This point is crucial, not only to forecasts based on international comparisons, but to other types of forecasting.

Some of these other methods also concentrate on the occupational distribution of the labor force, but use trend projections based on the country's own experience instead of international comparisons. In the absence of information about trends, they may adopt assumptions or norms that determine what proportion of the labor force should be employed in a particular occupation or should have a particular level of education. These trends or norms can then be combined with demographic projections of the labor force to give an estimate of the number of engineers, for example; the assumption here is that a constant proportion of engineers is employed in each industry or that a constant engineer-technician ratio exists.

Other methods of forecasting concentrate on the ratio between one type of manpower and a particular population parameter, such as the total labor force or the school population. For example, both forecasts of demand for teachers based on teacher-pupil ratios and forecasts for doctors based on doctor-population ratios rely on demographic forecasts combined with staffing norms. This method makes assumptions about the ratio between a particular occupation or educational category (for example, trained teachers) and the population. Like other forecasting methods, it assumes that different types of manpower cannot be substituted for one another (nurses for doctors, for example), and that capital cannot be substituted for manpower.

In the case of doctors or teachers, many of the objections to forecasting seem to have less force than in other occupations. For one thing, the link between educational qualifications and the job is closer than in many other occupations since educational or training requirements are usually specified in official regulations. In addition, the opportunities for substituting capital for teachers are limited, although television or broadcasting may be used to supplement the work of trained teachers. Moreover, since the government is usually the main, if not the only, employer of teachers (as well as the main source of supply of trained teachers), the need for forecasts is obvious, and since the training period is fairly long, the problem of time lags may also be severe. For all these reasons, forecasting teacher requirements is usually regarded as a fundamental prerequisite of any decision on education projects, and even countries that do not favor manpower forecasting as a general technique attempt to forecast teacher numbers.

Experience with teacher forecasts shows that even these are frequently inaccurate, however. Governments do face choices between different combinations of trained and untrained teachers, so that the scope for substitution and flexibility is far from negligible. These choices will be influenced by wage and salary levels, which also help to determine factors such as the rate at which teachers leave the profession. Actual teacher-pupil ratios frequently differ markedly from official norms, and the teacher-pupil ratio itself is influenced by factors such as the average size of class (which may vary between age groups, subjects, or type of school), the average teaching load per teacher, the extent of part-time teaching, or double-shift arrangements. The effects of all these factors on teacher demand and supply, together with the problems of planning teacher numbers, have been thoroughly reviewed by Williams (1979), who concludes that the planning of teacher supply should not be confined to forecasting numbers and determining annual intakes to teacher training colleges, but that it should seek to maintain flexibility so that teacher supply can adjust to rapidly changing situations.

In other words, even in the teaching profession, the assumption of fixed relationships can be challenged, and the accuracy of long-term forecasts based on simple teacher-pupil ratios questioned. In general, the problems of forecasting manpower on the basis of any labor-population ratios can be summarized as follows:

- The accuracy of demographic forecasts, particularly in developing countries, is often highly questionable.
- Many special factors need to be taken into account. For example, forecasts of the school population depend on wastage, repetition, the proportion of the age group choosing to enroll in voluntary education, and other factors that must be analyzed separately, although information is often scanty or inaccurate. Labor force projections depend on factors, such as the labor force participation of women, which may change.
- Technological change may cause a substitution of capital for labor or one kind of manpower for another.
- Changes in relative prices or wages may also cause substitution and may change the balance between engineers and technicians, doctors and nurses, or trained and untrained teachers.

Although labor-population ratios appear to be a simple way of forecasting manpower requirements, the accuracy of such forecasts will depend on the analysis of a wide range of factors, and the very notion of requirements assumes rigidities in the employment patterns that may not exist.

We come finally to a technique that has dominated the literature on manpower forecasting and that has been widely used in one form or another in many developed and developing countries—the projection of manpower-output ratios. This method was adopted in one of the earliest attempts to use manpower forecasting as a basis for educational planning, the Mediterranean Regional Project (MRP), which attempted to forecast manpower requirements in six Mediterranean countries: Greece, Italy, Portugal, Spain, Turkey, and Yugoslavia (OECD 1965). This project had considerable influence on educational planning, and the OECD later attempted to apply a similar methodology in Latin America (OECD

1967a, b). It is also of considerable interest because a technical evaluation of the project examined both the assumptions and the accuracy of the MRP forecasts (Hollister 1967).

In a series of ten stages, the MRP used output targets to derive manpower and education targets for each industry or sector. It employed a variety of forecasting methods, including employers' estimates, international comparisons, and extrapolation of input-output trends within the countries. In many cases, data were inadequate or unreliable, were not sufficiently detailed, or existed only for very broad sectors of the economy or occupational groups. Such problems are common to all forms of economic analysis or forecasting, however. What makes this type of exercise unusual is the crucial importance of the assumptions made about educational and occupational patterns. Each stage of the forecasting process is based on an assumption of certain fixed relationships—for example, between labor productivity and the occupational distribution or between occupation and education.

In his evaluation of the MRP exercise, Hollister (1967) subjected all these assumptions to sensitivity analysis, which showed how sensitive the final forecasts are to changes in key assumptions. He concluded that

- The concept of fixed manpower-output coefficients is misleading, and experience suggests that substitutability of manpower may be highly significant.
- The idea that there is a rigid relationship between occupation and education is the weakest link in the manpower requirements approach.
- Technological change is difficult to predict, but is very important in determining the occupational distribution of the labor force.
- Single-valued forecasts that make no attempt to measure the effect of alternative assumptions may be misleading for educational policy, which should be flexible enough to allow for uncertainties about the rate of technical progress and the link between formal education and the skills of workers.

The Links between Manpower, Levels of Output, and Education

Experience with all methods of manpower forecasting casts doubt on the validity of the whole notion of manpower requirements for meeting economic targets, which rests on "the rather rigorous link that has been assumed between productivity levels and occupational structure on the one hand, and between occupation and educational qualifications on the other" (Parnes 1962, p. 51). These relationships can not possibly be completely fixed at all times. A given output of textiles, for example, can be produced by using a large number of workers operating hand looms at home or a smaller number of workers and more capital equipment in a factory. The occupational structure of the textile industry would be quite different in these two cases.

Various studies have analyzed the extent of substitutability between manpower and capital or between different categories of manpower (Dougherty 1972; Psacharopoulos and Hinchliffe 1972). The extent to which one factor can be substituted for another (capital for labor or unskilled workers for skilled labor) is measured by the elasticity of substitution. If it is relatively easy to substitute one factor for another, the elasticity of substitution is high. If substitution is not possible, then the elasticity is zero, and the pattern of inputs is fixed. The assumption underlying the notion of manpower requirements is, as already emphasized, that the elasticity of substitution for educated manpower is zero or at least low.

There is now considerable evidence that substitution is both possible and significant, yet many manpower forecasts continue to assume fixed input-output coefficients. To calculate alternative estimates of manpower requirements on the basis of different combinations of inputs would greatly increase the complexity, not to mention the cost, of forecasting. If alternatives are ignored, on the other hand, a spurious sense of precision enters into the forecast of occupational distributions.

The situation is perhaps even more serious in the case of the link between education and occupation. Ample evidence now indicates that the relationship between education and occupation in developing countries is often far from rigorous, as forecasts assume; rather, it is fluid and changing over time. There are many reasons for this. One is that an increase in the supply of educated people may cause educational upgrading, so that the qualifications of new entrants will be higher than the qualifications of older workers. In addition, employers may regard on-the-job training or work experience as a substitute for formal education in equipping workers with necessary skills. Table 4-3 shows how the proportion of professional workers with a university degree varied among Latin American countries at the same point of time and how much change occurred in the same country between 1960 and 1970. Yet manpower forecasts frequently assume that this proportion is constant over ten or fifteen years, and that the proportion observed in one country can be applied to another.

Table 4-3. *The Changing Educational Composition of One Occupational Group in Selected Countries*

| Country | Percentage of professional workers with a university degree | |
	1960	1970
Panama	31.4	42.0
Paraguay	49.2	37.7
Costa Rica	52.4	18.1
Chile	28.5	24.4
Colombia	23.0	29.9

Source: Debeauvais and Psacharopoulos (forthcoming).

Evaluations of Manpower Forecasting: The Usefulness of Forecasts

In view of the problems already discussed and the uncertainties surrounding technical change and factors affecting economic growth targets, it is not surprising to discover that many forecasts of manpower requirements have proved inaccurate and unreliable. These results are documented in several assessments of manpower forecasts, among them a review of more than thirty manpower plans in Africa from 1960 to 1972 (Jolly and Colclough 1972) and a postmortem of similar forecasts in both developed countries (Canada, France, Sweden, the United Kingdom, and the United States) and developing countries (India, Nigeria, and Thailand) (Ahamad and Blaug 1973).

Jolly and Colclough concluded that most of the African manpower plans overestimated the growth of demand for educated manpower, and in some cases they overestimated future skilled manpower requirements by 100 percent. Ahamad and Blaug also found evidence of considerable inaccuracies in forecasts, and concluded that there were major weaknesses in the way manpower forecasts had been attempted and interpreted.

Two recent surveys (Snodgrass 1979; Debeauvais and Psacharopoulos forthcoming) have reiterated strong doubts about the accuracy and reliability of manpower forecasts, but have noted that in spite of previous adverse evaluations, manpower forecasting continues unabated. Their main criticisms of past forecasts are that they concentrate on the formal sector and neglect self-employment. They also neglect occupational mobility (the greater the extent of mobility, the less accurate and useful the forecast). Another important point is that countries at similar levels of economic development have experienced diverse educational and occupational structures. Finally, the longer the time horizon of the forecast, the less reliable it is likely to be.

In view of this dismal record, why, then, have manpower planners persisted in making manpower forecasts? One reason is that manpower requirements appear to be derived from an economic imperative tied to explicit national economic goals; they also appear to be simple and to provide a single method that can be applied to all categories of manpower from graduate physicists to bricklayers. The planner can concentrate on quantities, rather than on relative prices. Finally, because single-valued forecasts appear to give precise numerical targets, they are much more attractive to policymakers than cost-benefit analysis or other techniques that simply indicate "directions of change."

If one adds to this list the political pressure in many countries to plan for the replacement of expatriates at the time of independence, and the widespread fear that a shortage of qualified manpower would act as a bottleneck, it is hardly surprising that manpower forecasting so quickly became the main method of justifying educational expansion and investment.

For all their seeming precision and simplicity, however, the fact remains that such techniques are unreliable and may lead to mistaken judgments about investment priorities if they are used as the sole basis for assessment. A principal weakness is that manpower forecasts frequently neglect the question of costs, or at least the question of the relative cost-effectiveness of alternatives. The very idea of manpower *requirements* suggests necessity, rather than a choice between alternative ways of achieving the same level of output. Such forecasts also ignore the influence of relative prices in determining choices between alternative techniques and combinations of inputs.

The overall conclusion from twenty years' experience of forecasting is that manpower projections are not bad in themselves if they are viewed realistically—that is, if it is recognized that they are subject to wide margins of error and do not reflect hard and fast requirements of economic growth. Even so, experience in this field suggests that "most policymakers will not be able to take such a realistic view and for them the method appears to promise more than it can actually deliver" (Hollister 1983, p. 40).

In other words, the fault may lie not so much in the techniques of manpower forecasts, but in their interpretation. Statements of the manpower implications of economic targets are too often regarded by educational planners as necessary and sufficient conditions for the achievement of those targets. This is why many now advocate that manpower planners set aside the crystal ball of single-valued forecasts and the notion of requirements and begin to analyze the effects of alternative assumptions and the implications of alternative patterns of utilization; and that they introduce opportunity costs as a major criterion for choosing between alternatives (see, for example, Ahamad and Blaug 1973; Psacharopoulos 1984b). Such an approach would involve analysis of the economic concept of market *demand* for manpower (at a given or assumed wage rate), rather than the technological concept of manpower *requirements* or needs that has tended to dominate manpower planning in the past.

The Analysis of Manpower Demand and Selection of Educational Investments

In the past the World Bank has relied on forecasts of manpower requirements, or needs, to assess projects even though single-valued forecasts of requirements, by themselves, have seldom been regarded as sufficient justification for investment in education. The fact that Bank lending for education was initially heavily weighted in favor of higher education and professional and vocational education reflected the emphasis on these levels in the manpower plans of most developing countries. Requests for technical assistance, both from the World Bank and from other agencies, often took the form of requests for help in the preparation of manpower plans and forecasts. In the World Bank, as in many developing countries, there was substantial emphasis in the 1960s and early 1970s on the idea that educational investment should aim to satisfy manpower requirements. Now, however, the Bank is undergoing a change of direction, which can be seen in a number of ways.

The Education Sector Policy Paper of 1980 frankly admitted the problems of manpower forecasting in the past and argued that more emphasis should be placed on costs and the relationship between costs and benefits and costs and effectiveness (World Bank 1980). Similarly, a 1978 review of World Bank operations in the Education Sector examined the accuracy of manpower projections produced by project appraisals and found many examples of inaccuracies: for example, such developments as the Korean and Nigerian economic booms or the relative stagnation of the Sierra Leone economy could not have been foreseen over such a long time span. The oil price rise of 1973–74 prompted a great many technicians and other workers to migrate to the Middle East from Pakistan, Bangladesh, Korea, Tunisia, and Morocco, with the result that an acute shortage of technically skilled workers developed in countries such as Bangladesh and Pakistan, where unemployment of such workers had previously been extensive.

There is now a growing realization in the World Bank that the analysis of investment projects should be based on a variety of indicators, and that while assessment of manpower needs is important, it should be accompanied by other tests of the economic benefits of investment. One of these is the social rate of return, already discussed in chapter 3. Another is the factors determining the private demand for different kinds of education (see chapter 5) and the supply of people with the necessary educational qualifications for higher education. A recent project completion report for Zambia noted, for example, that even though the demand for engineers in Zambia is high, only 50 percent of the places in the College of Engineering can be filled; there simply is not a sufficient number of students with the requisite background skills in science and mathematics. Further investment in high-level engineering education in Zambia would clearly not be justified in this situation, even though there may be a shortage of engineers. The problem must be tackled in another way.

An assumption underlying many manpower forecasts is that vocational education helps to satisfy manpower needs, whereas general or academic education does not. This notion was first attacked as the "vocational school fallacy in development planning" by Foster (1965) in his analysis of vocational and general education in Ghana.

Since that time considerable evidence has accumulated that in some countries vocational education may be less profitable as an investment than general education, since the costs of highly specific vocational education are high, whereas the benefits, measured in terms of earnings, are often low for vocational school graduates in comparison with graduates with a general education (Psacharopoulos 1982). Experience with some World Bank projects in vocational education, particularly in middle-level agricultural education, also suggests that past estimates of future manpower requirements were overoptimistic at the planning stage. In Korea, Chile, El Salvador, Senegal, Tunisia, Morocco, and Tanzania, for example, some projects designed to increase agricultural education were modified to reduce enrollment targets after the original appraisal; in other projects there was either underenrollment or unemployment of agricultural school graduates.

Another problem is that vocational school graduates may be less flexible than workers with a general education. The uncertainty surrounding manpower forecasts, particularly uncertainty about future technical change, means that flexibility in the labor force is extremely important. In Tanzania, for example, rural training centers financed through World Bank loans were converted by the government into Folk Development Colleges offering a broader educational program than farmer training, and the programs of the agricultural training institutes were modified so as to reduce the output but improve the versatility of graduates.

Although the advantages and disadvantages of vocational and general education are being reassessed, this does not mean that vocational education is now regarded as having a low priority, but simply that general education is being seen in a new light; education that helps to make workers more flexible and adaptable to new technologies is now considered a valuable way of contributing to manpower development just as much as education that prepares them for specific jobs. It also means that more attention is now being paid to the relative costs (as well as the relative benefits) of vocational and general education, since the cost of vocational education is often much higher than general education, and this may lead to lower social rates of return, as observed in several developing countries (Psacharopoulos 1982).

The way in which the concept of manpower shortages is analyzed has also changed. The manpower requirements approach defines a shortage of manpower in terms of an imbalance between projected manpower requirements and supply. Yet planners in several countries have been warning of manpower shortages when there is no evidence of a shortage as judged by trends in relative wages. Of course, as pointed out in chapter 3, wage rates cannot be regarded as totally reliable measures of relative scarcities or surpluses. Nevertheless, trends in relative wages in different occupations provide an important indication of demand and supply, and therefore investment appraisals for the World Bank now frequently use relative wages as well as manpower forecasts to test the notion of shortages of skilled manpower.

The combined effect of all these changes in World Bank practice is that now greater emphasis is placed on manpower *analysis* rather than on manpower *forecasting* in the assessment of educational investment and labor market trends. This is now apparent, for example, in the increasing importance attached to tracer studies that follow the careers of cohorts of school leavers or graduates. Such studies are designed to monitor the progress of these people in the labor market and judge the effects of their education on productivity or earnings. Before examining a tracer study in detail, however, we should look at an example of manpower analysis.

A Case Study of Manpower Analysis

The analysis of present and past patterns of manpower utilization may provide more information on the operation of the labor market than simple projections of past manpower trends, as is demonstrated by the Human Resource Study undertaken in the Dominican Republic in 1974 and 1975. This study not only projected manpower demand and supply, but it also collected information about the labor force by means of special employer and worker surveys that were designed to improve understanding of the mechanisms influencing demand and supply and to identify imperfections or inefficiencies in either the labor market or the education system. World Bank staff members were involved in the collection and analysis of data for this study, and one of those who took part has written a useful case study that examines the value and the limitations of manpower surveys in the decisionmaking process (Dominguez 1979).

Experience shows that in many developing countries output targets tend to be overambitious, and the forecasts based on these targets inaccurate. The exercise in the Dominican Republic illustrates the danger of basing projections of demand on an unrealistic rate of growth. It also shows that forecasts of the balance between demand and supply depend on the accuracy of projections of the supply of and demand for qualified manpower. High rates of wastage and repetition in the Dominican Republic meant that forecasts of both supply and demand were likely to be inaccurate.

The study's estimates of manpower demand and supply in the Dominican Republic (see table 4-4) projected a surplus in many categories of university-educated professionals, but a shortage of other professional groups. In addition, the occupational pattern in different sectors was expected to remain stable. Further analysis of the data showed considerable differences in the educational qualifications of workers in the same occupational group. The forecasts also took no account of relative wages and salaries. In other words, the projected shortages and surpluses assumed no change in relative salaries, standards of service, or occupational distinctions. Yet the fact that there is a projected surplus of doctors and medical technicians, combined with a shortage of nurses, suggests that existing patterns of work and relative wages may change in response to the supply of people with different medical qualifications. The demand for workers with medical qualifications also depends on government policy with respect to the provision of health services. Any decision to improve standards of health care would change the demand for all these occupational groups; similarly, changes in work patterns might cause a substitution of doctors for medical technicians, or vice versa.

This example is instructive because it shows that projections of the balance between demand and supply should be taken as the starting point for further analysis, rather than the final statement of manpower requirements. Thus, when questions were posed about the likely effect of changes in work patterns or relative wages, it became necessary to collect additional information about the labor market in the Dominican Republic, some of which was obtained through special analysis of certain occupations, for example, medical personnel. The type of information that might be collected by means of special surveys includes data on

Table 4-4. *Supply of and Demand for Labor Force, by Occupation, Dominican Republic*

Occupation	Stock 1970	Demand 1985	Supply 1985	Surplus or deficit
Professionals and technicians with university education				
Engineers and architects	1,415	2,106	10,572	8,466
Economists, sociologists, and mathematicians	209	439	8,056	7,617
Accountants, assistant accountants	1,381	2,165	7,592	5,427
Paramedics, medical technicians	294	806	4,552	3,746
Doctors, surgeons, and dentists	1,604	4,135	6,636	2,501
Agronomists and veterinarians	491	1,152	2,329	1,177
Chemists, physicists, and pharmacists	764	2,069	3,108	1,039
Lawyers and judges	1,260	1,660	2,559	899
Subtotal	7,418	14,532	45,404	30,872
Professionals without university education	1,106	5,304	4,598[b]	−706
Professionals with some university education	796	2,286	1,519[b]	−767
Industrial technicians	3,742	7,793	3,292[b]	−4,501
Professors[a]	5,603	14,789	12,078[b]	−2,711
Nurses, midwives	2,889	7,515	4,497	−3,018
Subtotal	14,136	37,687	25,984	−11,703
Managers and administrators	4,548	9,958	4,000	n.a.
Other teachers	8,207	21,685	n.a.	n.a.
Office workers	69,453	121,456	348,000	n.a.
Retailers	52,954	86,819	n.a.	n.a.
Subtotal	135,162	239,918	352,000	n.a.
Manual laborers				
Semiskilled and unskilled laborers[c]	642,408	1,155,937	1,303,000	n.a.
Other mechanics	1,577	4,948	14,830	9,882
Subtotal	643,985	1,160,885	1,317,830	n.a.
Skilled workers				
Electricians	3,878	9,725	8,239	−1,486
Plumbers and metal workers	7,665	8,737	4,839	−3,898
Automobile mechanics	14,054	36,215	14,657	−21,558
Subtotal	25,597	54,677	27,735	−26,942
Total[d]	826,298	1,507,699	1,768,953	261,254

n.a. Not available.

a. Including university professors and primary and secondary teachers.

b. Excess demand will be met by the surplus of professionals in other occupational categories.

c. Including farm workers, miners, day laborers, and the like.

d. Excluding military personnel.

Source: Dominguez (1979), p. 8.

earnings, hiring practices of employers, the extent to which work experience is regarded as a substitute for formal education, the availability of training, and the extent of job mobility.

One special study in the Dominican Republic was designed to collect data on earnings. Because education in this country has expanded rapidly in recent years, younger workers have higher levels of education than older workers. The survey showed, however, that older workers with less education earned as much as younger workers with more education but less work experience. This suggests that in the Dominican Republic, at least, education and work experience are regarded by employers as substitutable, and that present employment patterns reflect past supply and cannot be taken as a reliable indicator of future demand.

The special employment survey provided other insights into the workings of the labor market. For example, young workers with higher education stayed unemployed longer than other young workers, apparently because they were more selective and preferred to explore the labor market for better job-entry opportunities. In addition, there was a closer relationship between earnings and occupation than between earnings and level of education. Despite their low educational level, self-employed workers frequently earn more than employees with more education, and, according to Dominguez, this represents a payoff to risk and entrepreneurial activity. Finally, the survey suggested that the Dominican labor market is rigid and hierarchical, and offers few opportunities for promotion through on-the-job training; less educated workers feel themselves to be trapped in low-status, low-salary jobs with few prospects of mobility and believe that it is the lack of a formal education that prevents them from moving to better-paid jobs.

The detailed analysis of the labor market suggested that the Dominican Republic does not have one homogeneous labor market, but different conditions for different locations, types of firm, and categories of worker. Rate-of-return calculations showed that there were considerable differences between the rates of return for men and women in the two main cities, as shown in table 4-5. Returns are consistently higher in Santiago

Table 4-5. *Rates of Return on Education*
(percent)

Level	Place	Males	Females
Secondary	Santo Domingo	8.3	5.2
	Santiago	11.1	19.2
Higher	Santo Domingo	9.4	4.7
	Santiago	23.6	13.5

Source: Dominguez (1979).

than in Santo Domingo, and whereas men have a higher rate of return to secondary education than women in Santo Domingo, the reverse is true in Santiago. In this situation an average rate of return for both sexes and the country as a whole would be misleading. Similar findings in other countries of differences in the labor market for different categories of workers have led some economists to talk of dual labor markets or segmented labor markets. This concept has aroused considerable controversy in both developed and developing countries, and reviews of the literature (for example, Carnoy 1980; Psacharopoulos 1978) come to different conclusions about whether differentiation in the labor market implies the existence of distinct labor markets for different groups of workers. There is general agreement, however, that important barriers to mobility exist in many countries.

Dominguez explored the question of mobility through a survey of employers' hiring and promotion practices. He found not only a number of barriers to mobility, but also some other interesting results. For example, education appears to be the most important factor explaining promotions and increased earnings among professionals in the Dominican Republic, whereas honesty and loyalty are the most important criteria among blue-collar workers. Similarly, many firms appeared to attach greater weight to education when hiring professional workers, whereas letters of reference were given more weight in the case of blue-collar workers. Furthermore, a survey of starting salaries and present salaries among a sample of workers showed an interesting relationship between education, occupation, and salary progression. The fact that present salary was more highly correlated with education than starting salaries suggests that employers used educational level as a criterion for both promotion and hiring decisions.

The survey also included follow-up studies on recent graduates. It was found that 70 percent of university graduates were employed, but only 44 percent of secondary vocational school graduates. On the average, those who were employed had spent four months looking for their present job, but the waiting period varied for different occupations. Technicians, machine operators, metal workers, and architects found work in the shortest period, and most technicians found work within half a month, whereas medical technicians took an average of eight-and-a-half months to find employment. This kind of information is more useful as an indicator of demand for different skills than a manpower forecast that simply states there is a shortage of technicians.

Interesting information was also collected on the opinions of recent graduates about the vocational importance of education. Surprisingly, 40 percent of all the employed graduates believed that their job could be filled by someone with less education. This finding does not support the idea that educational requirements for jobs are rigid, and there is evidence of considerable differences in the educational attainment of professional workers. This is one indication of flexibility in the labor market; yet, other signs (such as little job mobility and little opportunity for on-the-job training) suggest a lack of flexibility.

The results of the labor market surveys therefore suggest a need for policies that will increase labor market flexibility and the mobility of workers, rather than rapidly expand higher education. The considerable differences between rates of return to secondary and higher education in different towns, the fact that many graduates considered themselves overeducated for their jobs, and the widespread belief among many university graduates that connections are more important than education in determining access to jobs—all suggest that there is no general shortage of university-level manpower in the Dominican Republic. This suspicion is confirmed by the forecasts of manpower demand and supply that project surpluses of many types of university-educated manpower. Yet the private demand for university places is very strong, and according to Dominguez (1979, p. 10), "university education in the Dominican Republic is now under pressure from an overwhelming number of applicants. Continued expansion may reduce quality, reduced efficiency may raise costs, and the resources required even to maintain present levels may strain Dominican resources."

This analysis of manpower patterns in the Dominican Republic shows how an examination of labor market characteristics may point to the need for measures that will promote flexibility and mobility (such as greater provision of informal training), and that will not rely too much on formal education. The analysis also focuses on the need to improve educational efficiency and on the importance of costs in determining investment priorities. Information from the special manpower surveys about the labor market and the relationship between education and earnings was more useful than a simple statement of projected demand and supply.

The World Bank is encouraging this trend in projects and research by emphasizing the development of tracer studies that would collect information about the progress and experience of small samples of labor force entrants or school or university graduates. It has also been proposed that manpower development planning units be established to monitor and analyze manpower trends. The function of these units would not be to make forecasts or projections, but to analyze, on a regular basis, the working of the labor market and the relations between education, employment, and the economy. The next two sections look briefly at these two developments.

The Use of Tracer Studies

The purpose of tracer or follow-up studies is, as the name implies, to examine the subsequent careers and employment of a sample of school leavers or graduates. Some are designed to compare the success with which different educational institutions prepare young people for employment, while others are more concerned with collecting information about the labor market. Although the purposes of particular tracer studies may differ, most seek information on the following types of questions:

- How do people learn about the programs they enter and why do they enter them?
- To what extent do schools help their students find work or advise them about additional education?
- How do graduates set about looking for work? How do they learn about opportunities? What is the range of work types and the minimum level of earnings graduates will accept, and does this change during a period of unemployment?
- How are people supported while they are looking for work?
- Is it getting harder or easier to find a particular type of work?
- How quickly do graduates find their first work and how mobile are they thereafter?
- What sort of work do graduates prefer and expect, and how does this relate to what they get?
- What sort of jobs do graduates enter or what sort of work do they do and what do they earn? How do they compare in these respects with other groups in the labor force?
- To what extent is the content of work obtained related to the type of education received? Why do some people not enter jobs for which they are trained?
- What are the obstacles to setting up in self-employment?
- Do the unemployed have no work because none is available or because they are waiting for a particular job or level of earnings?
- To what extent are nonschool factors such as family background, age, and location important in achieving success in school and in the labor market?
- What is the cost of producing one successful graduate, and how does this relate to the extra earnings of graduates?
- Which students want to take further education, and how many are successful?

Answers to these questions can then be used to compare the labor market performance of different types of graduate, for example, those from general academic programs and those from vocational schools, or those from diversified secondary schools and those from traditional schools. (The DiSCuS study discussed below is an example of such a study.) This information can be compared with the costs of the different programs; it can also throw light on the factors affecting occupational choice, mobility, and wastage, and it can be used to provide both students and employers with better labor market information. Because many countries are now placing greater emphasis on vocational guidance, the results of tracer studies can be useful in providing information about the labor market experience of school leavers not only in the formal urban sector, but also in the informal sectors, both urban and rural, which are often not represented in an official manpower survey or census.

Although the World Bank has promoted the use of tracer studies to assess the impact of education and the extent to which it satisfies manpower needs, a recent review of the research components in educational loans from the Bank showed that a large proportion of the tracer studies that have been agreed upon by the Bank and individual governments have not been undertaken or completed. The problem may be that the purposes, methodology, and usefulness of tracer studies are not well understood. The Bank has therefore produced a set of guidelines and suggestions regarding the design and operation of such studies (see Psacharopoulos and Hinchliffe 1983).

The main advantages of a tracer study that concentrates on today's graduating students and follows them over time is that it provides completely up-to-date information on the state of the labor market and on the behavior of students who have recently entered it. Such a study also captures information across the entire spectrum of employment opportunities, rather than concentrating (as establishment surveys often do) on those who end up in large firms. If tracer studies can be periodically repeated for new sets of graduating students, then it may be possible to discover trends in the labor market and to monitor the effectiveness of new institutions and new programs with respect to their graduates' futures.

An example of an ambitious tracer study is the recent analysis known as DiSCuS, which examined the diversified curriculum introduced in secondary schools in Colombia and Tanzania (Psacharopoulos and Loxley forthcoming). This was a longitudinal tracer study to collect information on students at three points in time: while still in school, one year after graduation, and several years after graduation. So far, information has been collected on a cohort of Colombian and Tanzanian graduates one year after they left school and a cohort of Colombian graduates after three years.

Such a survey can yield valuable information on many issues connected with manpower development and labor market characteristics, and it can compare this information with cost data. The DiSCuS study has estimated private and social rates of return to different types of schools, and it has carried out achievement tests, the results of which can be used in cost-effectiveness comparisons (see chapter 8). It also collected information about employment, the incidence of job search, and future career aspirations of graduates. Such data may be useful in evaluating a large investment, such as fundamental curriculum reform. Because it is costly to collect all this information, the value of the data must be carefully weighed against the costs of collection.

The costs will depend on many factors, such as how many graduates are interviewed. In a tracer study of secondary school leavers in Zambia in the early and mid-1970s, 66 percent of the student sample were contacted one year after graduation and 50 percent at the end of the next year. In an Indonesian study, however, the percentages were higher, 89 and 66 percent, respectively. Because of the long period before the results of the tracer study are available, an alternative to the follow-up approach that has sometimes been adopted is a retrospective tracer study, which attempts to discover the postschool experiences of graduates by asking them to recount their educational, unemployment, employment, and earnings records since graduation. The Colombian DiSCuS project included such a retrospective tracer study (Psacharopoulos and Zabalza 1984). A more ambitious approach would be to contact sets of graduates from different years to identify past trends. One of the most successful studies of this type analyzed the experiences of students graduating from a technical university in Malaysia in each of the years between 1967 and 1978 (Institut Teknologi Mara 1980). A number of tracer studies have now been successfully carried out in developing countries, for example, in Chile (Schiefelbein and Farrell 1982), Egypt (Sanyal 1982), and Swaziland (Sullivan 1981). Before the results of such studies can be exploited fully, however, countries will need to develop the machinery to analyze feedback from labor market surveys regularly, an example of which would be the manpower development planning units mentioned earlier.

The Role of a Manpower Planning Development Unit

In the past, the preparation of a manpower plan, or forecast, was often regarded as one step in the formulation of a long-term economic plan, and therefore it was expected to be repeated at intervals of about four or five years, rather than handled as a regular, ongoing activity. The types of manpower analysis described in this chapter, however, require regular feedback, updating, continuous analysis, and monitoring, rather than the preparation of occasional quantitative projections. In a recent review of manpower planning issues, Dougherty (1983) has suggested that manpower planning units in developing countries should set up an annual routine for obtaining continuous information on the operation of the labor market and the education and training systems, and for feeding this information back to employers, graduates, and educational institutions.

Such an annual review would include quantitative projections for occupations, such as teachers or health personnel, for which demand can be related to demographic, social, or political norms. The projections should be regularly updated on the basis of the latest information on wastage, wage and salary trends, and other indicators of actual demand and supply conditions, rather than fixed ratios or norms.

In addition, the annual review would collect and disseminate a wide range of information, including not only employment and unemployment statistics, but also earnings data, information on hiring practices, the amount of training provided by employers, job turnover and wastage rates, job vacancies, and education statistics (for example, enrollment and graduation rates by type of institution, current and capital costs of educational institutions, and subsequent destinations of graduates). Much of this information could be collected by sample surveys rather than a large-scale manpower census; it could use the experience of key informants, such as the principal employers in either the public or private sector, whose subjective assessments of the current state of the market for different occupational groups might supplement more general statistics. Dougherty warns against a policy of simply collecting a mass of data, however; in particular, he suggests that the temptation to establish a manpower data bank, which tends to be very expensive to maintain, should be resisted. Instead, the emphasis should be on analyzing the data regularly so as to reveal trends in manpower utilization and the relationships between education and employment, and to disseminate information among institutions and individuals on a regular basis.

The difference between a manpower planning unit that carries out this type of analysis and a conventional manpower unit is that the former emphasizes manpower planning as a continuous process rather than a technique, whereas the latter is engaged primarily in preparing quantitative projections for the overall development plan.

Manpower Planning as a Continuous Process

Whatever the weaknesses of manpower forecasting as an approach to planning educational investment, the fact remains that manpower development is still the dominant objective of many educational plans or project proposals. Despite the inaccuracies of past forecasts, governments need to prepare forecasts of demand for and supply of professional workers in the public sector, which, as noted earlier, is the principal employer of highly qualified manpower in many developing countries. Moreover, estimates of future manpower needs have become more or less a requirement for securing budgetary funds and international agency loans for educational expansion. The result is that during the 1970s, at the same time that the reviews of manpower forecasting described in this chapter reached their skeptical and negative conclusions, manpower forecasting became more popular than ever. It would clearly be unrealistic to argue that such attempts to forecast manpower demand should be abandoned, and that is not what is being proposed here.

The main lesson that arises from experience within the World Bank and in other agencies, as well as in individual developing countries, is that manpower planning should be regarded as a much broader activity than manpower forecasting. The idea of manpower *requirements* should be abandoned, however, and instead, manpower planning should be concerned with patterns of manpower *utilization* and with the *demand* for manpower, in the sense of the numbers of workers who will be employed at a given wage rate.

It should also be closely concerned with evaluating costs and comparing alternatives. A recent assessment of current manpower planning in developing countries (Psacharopoulos 1984b) advocates more reliance in the future on a greater variety of signals for assessing educational and training priorities. (Some of the signals that may be useful have already been discussed here and in chapter 3, while chapter 5 looks at another important signal, the private demand for education.)

The implication of this new approach to manpower planning is that less emphasis should be placed on planning *techniques* and more on the concept of planning as a *continuous process*. This process should encompass analysis of the consequences of alternative policies in terms of demand and supply for different types of manpower, alternative ways of producing skills (including informal training as well as formal education), comparisons of their relative costs, and analysis of alternative ways of using manpower that investigates the scope for substitutability and flexibility in the labor market and the role of wage and other financial incentives in determining demand for education and for educated manpower.

Some of the examples cited thus far indicate that manpower studies are beginning to pay more attention to actual labor market conditions in developing countries, rather than searching for general patterns that can be used for forecasting. Another example that could be added to this group is a useful discussion of methods of planning teacher supply and demand (Williams 1979), which argues that planning teacher supply should involve far more than forecasting and determining annual intakes to training colleges; it should be concerned with the wide range of alternative sources of recruitment that can be tapped, and the factors affecting the retention and utilization of teachers. Above all, it should seek to maintain flexibility in teacher supply and increase its capacity to respond to changed circumstances. A similar emphasis on present patterns of recruitment and utilization is evident in a recent series of labor market studies carried out by the International Institute for Educational Planning (IIEP) in Paris (1980) that investigated the labor market for educated manpower in Panama, Indonesia, and Kenya and the links between education and self-employment. The final conclusion of these studies was that "even when we have gained a more thorough knowledge of the social context and, in particular, of the interactions between education and employment, even when the objectives of the educational and productive systems have been fixed, and even when society's needs for skilled manpower have been estimated with the greatest possible precision, we will still have to accept discontinuities between future projections and real events" (IIEP 1980).

This comment suggests a continuing but limited role for forecasting. If manpower forecasting is to play a more subsidiary role, however, manpower analysis in the wider sense, as advocated in this chapter, will still be a necessary ingredient of investment decisions. Such manpower analysis may include forecasts for particular occupations, which experience has shown may be more accurate than for the economy as a whole. It should also include elements of cost-benefit analysis and analysis of internal efficiency and cost-effectiveness. The increased emphasis on primary schooling, on investments that improve the quality of education, or on the geographical or social distribution of educational opportunities, rather than on quantitative expansion, and the growing concern about cost constraints in developing countries mean that manpower development can no longer claim a preeminent role in guiding investment policies.

Dougherty (1983) argues that because traditional manpower forecasting is subject to large margins of error and because the newer forms of manpower analysis do not pretend to yield precise indicators, manpower development decisions must be taken on the basis of informed conviction rather than in response to projections of requirements. The conclusion of this chapter is that the informed conviction on which manpower decisions should be made ought to be based on an analysis of actual manpower patterns along with accurate information about the labor market, wage rates, hiring practices, and all the factors that determine the utilization of manpower. It should also be based on an analysis of costs and finance, internal efficiency, and the determinants of individual demand for education.

References

Ahamad, B., and M. Blaug. 1973. *The Practice of Manpower Forecasting: A Collection of Case Studies*. Amsterdam: Elsevier.

Ashby, E. 1960. *Investment in Education: The Report of the Commission on Post-School Certificate and Higher Education in Nigeria*. Lagos: Federal Ministry of Education.

Blaug, Mark. 1967. A Cost-Benefit Approach to Educational Planning in Developing Countries. Economics Department Report no. EC-157. Washington, D.C.: World Bank.

Carnoy, M. 1980. Segmented Labour Markets. In *Education, Work and Employment*. Vol. 2. Paris: International Institute for Educational Planning.

Debeauvais, M., and G. Psacharopoulos. Forthcoming. Forecasting the Needs for Qualified Manpower: Towards an Evaluation. In *The Practice of Manpower Planning Revisited*, ed. K. Hinchliffe and R. Youdi. Paris: International Institute for Educational Planning.

Dominguez, J. 1979. Forecasting Manpower Requirements: The Case of the Dominican Republic. Case Study and Exercise Series. Washington, D.C.: World Bank, Economic Development Institute.

Dougherty, C. R. S. 1972. Substitution and the Structure of the Labour Force. *Economic Journal* 82 (March):170–82.

————. 1983. Manpower Planning from Three Points of View: Country, Technical Assistance Agency and Lending Agency. In *Manpower Issues in Educational Investments*, ed. G. Psacharopoulos et al. World Bank Staff Working Paper no. 624. Washington, D.C.

Foster, P. J. 1965. The Vocational School Fallacy in Development Planning. In *Education and Economic Development*, ed. C. A. Anderson and M. J. Bowman. Chicago: Aldine.

Freeman, Richard. 1971. *The Market for College-Trained Manpower: A Study in the Economics of Career Choice*. Cambridge, Mass.: Harvard University Press.

Hinchliffe, Keith. 1973. Nigeria. In *The Practice of Manpower Forecasting: A Collection of Case Studies*, ed. B. Ahamad and M. Blaug. Amsterdam: Elsevier.

Hollister, Robinson. 1964. The Economics of Manpower Forecasting. *International Labour Review* 89, no. 4 (April):371–97.

————. 1967. *Technical Evaluation of the First Stage of the Mediterranean Regional Project*. Paris: Organisation for Economic Co-operation and Development.

————. 1983. A Perspective on the Role of Manpower Analysis and Planning in Developing Countries. In *Manpower Issues in Educational Investment*, ed. G. Psacharopoulos et al. World Bank Staff Working Paper no. 624. Washington, D.C.

Horowitz, M. A., M. Zymelman, and I. L. Herrnstadt. 1966. *Manpower Requirements for Planning: An International Comparisons Approach*. 2 vols. Boston, Mass.: Northeastern University.

International Institute for Educational Planning (IIEP). 1980. *Education, Work and Employment*. Vols. 1–2. Paris.

International Labour Organisation (ILO). 1979. *Manpower Assessment and Planning Projects in Asia: Situation, Problems and Outlook*. Geneva.

Institut Teknologi Mara. 1980. *Student Tracer Study: Institut Teknologi Mara*. Kuala Lumpur.

Jolly, R., and C. Colclough. 1972. African Manpower Plans: An Evaluation. *International Labour Review* 106, nos. 2–3 (August–September):207–64.

Layard, R., and J. Saigal. 1966. Educational and Occupational Characteristics of Manpower: An International Comparison. *British Journal of Industrial Relations* (July):222–67.

Netherlands Economic Institute. 1966. *The Educational Structure of the Labor Force*. Rotterdam.

Organisation for Economic Co-operation and Development (OECD). 1965. *The Mediterranean Regional Project: An Experiment in Planning by Six Countries*. Paris.

————. 1967a. *Education, Human Resources and Development in Argentina*. Paris.

————. 1967b. *Problems of Human Resources Planning in Latin America*. Paris.

————. 1970. *Occupational and Educational Structures of the Labour Force and Levels of Economic Development: Possibilities and Limitations of an International Comparison Approach*. Paris.

Parnes, Herbert. 1962. Planning Education for Economic and Social Development. In *Forecasting Educational Needs for Economic and Social Development*, ed. H. Parnes. Paris: OECD.

Psacharopoulos, G. 1978. Labour Market Duality and Income Distribution: The Case of the United Kingdom. In *Personal Income Distribution*, ed. W. Krelle and A. F. Shorrocks. *Proceedings of a Conference of the International Economic Association*. Amsterdam: North-Holland.

————. 1982. The Economics of Higher Education in Developing Countries. *Comparative Education Review* 26, no. 2 (June):139–59.

————. 1984a. Indonesia: Manpower Considerations in the Energy Sector. Washington, D.C.: World Bank, Education Department.

————. 1984b. On the Assessment of Training Priorities in Developing Countries: Current Practice and Possible Alternatives. *International Labour Review* (September–October).

Psacharopoulos, George, and Keith Hinchliffe. 1972. Further Evidence on the Elasticity of Substitution among Different Types of Labor. *Journal of Political Economy* 80(4):786–92.

———. 1983. Tracer Study Guidelines. Washington, D.C.: World Bank, Education Department.

Psacharopoulos, George, and William Loxley. Forthcoming. *Diversified Secondary Education and Development: Evidence from Colombia and Tanzania.* Baltimore, Md.: Johns Hopkins University Press.

Psacharopoulos, George, and Antonio Zabalza. 1984. *The Destination and Early Career Performance of Secondary School Graduates in Colombia.* World Bank Staff Working Paper no. 653. Washington, D.C.

Psacharopoulos, G., K. Hinchliffe, C. Dougherty, and R. Hollister. 1983. *Manpower Issues in Educational Investments.* World Bank Staff Working Paper no. 624. Washington, D.C.

Sanyal, B. 1982. *University Education and the Labor Market in the Arab Republic of Egypt.* Oxford: Pergamon.

Schiefelbein, E., and J. P. Farrell. 1982. *Eight Years of Their Lives: Through Schooling to the Labor Market in Chile.* Ottawa: International Development Research Centre.

Snodgrass, D., with D. Sen. 1979. Manpower Planning Analysis in Developing Countries: The State of the Art. Development Discussion Paper no. 64. Cambridge, Mass.: Harvard Institute of International Development.

Sullivan, G. 1981. *From School to Work: Report on the School Leaver Tracer Project (Swaziland).* Oxford: Cotswold Press.

Unesco. 1968. *Educational Planning: A Survey of Problems and Prospects.* Paris.

Williams, Peter. 1979. *Planning Teacher Demand and Supply.* Paris: Unesco/ International Institute for Educational Planning.

World Bank. 1980. *Education.* Sector Policy Paper. Washington, D.C.

Zymelman, Manuel. 1980a. *Occupational Structures of Industries.* Washington, D.C.: World Bank, Education Department.

———. 1980b. *Forecasting Manpower Demand.* Washington, D.C.: World Bank, Education Department.

5

The Private Demand for Education

Educational investment, whether based on cost-benefit analysis, forecasts of manpower demand, or other criteria, cannot be adequately assessed unless estimates of future demand for education and student numbers are taken into account. The number of pupils or students enrolled in education is determined by a variety of economic and noneconomic factors. Obviously, government policy on the supply of places and the allocation of funds for education has an important influence on demand, since it determines the level of fees and the level of financial support for students (through scholarships, grants, or loans). Social factors and attitudes are also important, however.

Because the terms *social demand* and *private demand* are often used interchangeably, confusion sometimes arises in discussions of demand. The total number of pupils or students enrolled in an education system is the result of a series of private investment decisions. Together, however, these private decisions constitute social demand. It has been suggested in some developed countries that social demand should be the criterion for educational investment decisions. In the United Kingdom, for example, the Robbins Committee on Higher Education (1963) rejected both manpower forecasting and cost-benefit analysis as a reliable guide for decisions about the scale of higher education, but claimed "as an axiom that courses of higher education should be available for all those who are qualified by ability and attainment to pursue them and who wish to do so" (p. 8). Educational planning based on this axiom is widely known as the social demand approach. Its underlying rationale is that social investment should aim to satisfy *private* demand, and that the policymaker must therefore forecast future demand by taking into account all the economic and noneconomic factors that determine private demand for education. Enrollment projections are necessary, of course, whether investment decisions are based on forecasts of private demand, manpower demand, analysis of costs and benefits, or simply rough judgments about the cost-effectiveness and relative advantages of different projects. The point is that estimates of future enrollments need to take into account all the determinants of private demand for education as well as demographic trends if they are to be reasonably accurate. Furthermore, they must not overlook wastage or repetition.

One of the criteria by which the success of World Bank education projects is judged is the extent to which the enrollment targets adopted in project planning and appraisal are actually achieved. In a sample of education projects completed between 1972 and 1977 (see table 5-1), the original enrollment targets were revised downwards in many cases because of cost or changes in national policy. These revised targets were by and large achieved or exceeded—an outcome that reflects the strong private demand for school or higher education in these countries. In some cases, however, the demand for places was underestimated and overcrowding occurred as a result.

To forecast enrollments accurately, then, analysts need to consider three basic factors: demographic trends, which will provide accurate estimates of the school-age population; the determinants of private demand for education, that is, the factors that determine whether or not pupils or students choose to enroll in education; and promotion, repetition, and dropout, which will indicate how many of the pupils or students who originally enrolled will remain in the system and ultimately graduate. This chapter is mainly concerned with the second of these issues, the private demand for education, although it briefly examines recent analysis of demographic trends. Wastage and repetition, which are a serious problem in many developing countries, are discussed in chapter 8.

Demographic Analysis

The first step in projecting future school enrollments is to analyze existing demographic data on the age structure of the population, trends in birth and mortality rates, the natural rate of population growth, and the net reproduction rate. World Bank analysis of demographic data and projections of future population trends provide estimates of future population, by age group, for every country. (For more detailed information on demographic techniques, see Liu [1969] and Chau [1969].) According to Bank estimates, both birth and death rates will continue to fall in developing countries and population will continue to increase rapidly until well into the twenty-first century (figure 5-1). This forecast has many implications for investment in education.

The rapid increase in population in developing countries means that the school-age population is much higher as a proportion of the total

Table 5-1. *Achievement of Enrollment Targets in a Sample of Education Projects, Mid-1970s*

Project	Level	Original target	Modified target	Enroll-ment	Difference (percent)[a]
Cameroon I	Secondary	4,610	—	6,981	+51
	Subprofessional	1,220	—	2,331	+91
	Teacher training	1,490	—	455	−70
	Higher education	120	—	162	+35
Chile II	Subprofessional	38,000	—	41,052	+ 8
Chile III	Subprofessional	4,466	—	1,415	−68
	Teacher training	5,300	—	n.a.	n.a.
Colombia I	Secondary	48,070	—	26,966	−44
Colombia II	Secondary	31,000	—	14,409	−54
El Salvador	Secondary	19,250	—	8,744	−55
	Subprofessional	1,580	—	754	−52
Guyana I	Secondary	5,696	—	5,700	n.a.
	Teacher training	660	—	582	−18
Indonesia I	Subprofessional	13,800	—	11,220	−19
Indonesia II	Subprofessional	3,280	—	3,768	+15
Kenya I	Secondary	15,850	—	17,373	+10
	Subprofessional	2,930	—	3,631	+24
Kenya II	Subprofessional	3,720	—	4,044	+ 9
	Teacher training	5,640	—	6,021	+ 7
	Higher education	300	—	309	+ 3
Korea I	Subprofessional	34,140	—	36,879	+ 8
	Teacher training	2,220	—	1,519	−32
Morocco I	Secondary	25,679	—	26,584	+ 3
	Teacher training	3,996	—	n.a.	n.a.
Philippines I	Higher education	2,500	—	3,650	+46
Senegal I	Subprofessional	3,885	—	n.a.	n.a.
Sierra Leone I	Secondary	5,960	—	n.a.	n.a.
	Subprofessional	1,930	—	806	−58
	Teacher training	250	—	145	−42
Tunisia II	Secondary	12,390	—	16,824	+36
	Subprofessional	7,260	—	n.a.	n.a.
All projects	Secondary	162,545	149,395	123,581	−24 (−17)
	Subprofessional	75,042	69,100	72,473	− 3 (+ 5)
	Teacher training	15,560	8,460	8,722	−43 (+ 4)
	Higher education	2,920	2,920	4,121	+41 (+41)

n.a. Not applicable.

—Not applicable.

Note: The subprofessional category includes technical, vocational, and agricultural training.

a. Difference between original target and actual enrollment. Difference between modified target and actual enrollment is in parentheses.

Source: World Bank data.

Figure 5-1. *Trends in Birth and Death Rates, 1775–2050*
(births and deaths per 1,000 population)ᵃ

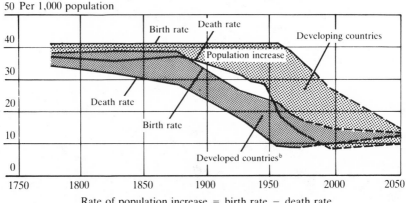

a. Crude birth and death rates. The projected increases in death rates after about 1980 reflect the rising proportion of older people in the population.
b. Include industrialized countries, the U.S.S.R., and Eastern Europe.
Source: World Bank (1980a), p. 64.

population than in developed countries, where the rate of population growth is lower. In 1973, 21 percent of the population in developing countries was of primary school age, compared with 10 percent in OECD countries. The demographic burden is therefore much greater in developing countries, and continued population growth will increase the demographic pressure on education (Zymelman 1979).

This demographic pressure is particularly striking in South Asia. In 1980 the population for the region as a whole was estimated at 896 million. A growth rate of 2.2 percent a year, on the average, would mean an additional 20 million individuals a year and a total population in the region of about 1.3 billion by 2000. Grawe (1980) has pointed out that the impact of this growth on demand for education and health care is equivalent to adding to the region another India populated exclusively by individuals of preschool or school age and those forming new families— the very ages when potential demand for maternal and child health services is greatest. The relative burdens of countries in this region can be seen in the projected increase in the primary-school-age population between 1980 and 2000 in India, Bangladesh and Pakistan (table 5-2). In India, the projected increase of 19 million represents slightly more than 20 percent of the 1980 primary school population, but in Bangladesh the increase of 5 million would mean an increase of nearly 40 percent, while

Table 5-2. *Projections of Primary School-Age Cohorts*

Country	Primary school population (aged 5–9, millions)		Growth rate (percent a year) 1980–2000
	1980	*2000*	
India	87	106	1.0
Bangladesh	13	18	1.6
Pakistan	13	21	2.4

Source: Grawe (1980), p. 104.

in Pakistan, an increase of 8 million would amount to an increase of more than 60 percent. In other developing regions the burden is even greater. In Africa, most countries would have to more than double total school enrollments by the end of the century simply to retain the low primary school enrollment ratios of 1980 (Davies 1980).

Considerable increases in the school-age population of developing countries will continue despite planned reductions in fertility rates. Nevertheless, declining fertility will help to reduce the demographic pressure on education. One of the factors that has contributed to a reduction in fertility in developing countries is in fact education. A review of the extensive research on the links between education and fertility (Cochrane 1979) has shown a consistent and well-documented inverse relation between the two. The relationship is a complex one, however, and increased education may, in some cases, increase fertility before reducing it. (The links between education and fertility are examined in chapter 10.) Adding to the complexity are other influences on fertility such as government programs that encourage family planning and a variety of socioeconomic factors (see figure 5-2). Thus there are many reasons why population forecasts may be inaccurate. For purposes of school enrollment projections, however, many of the births that will determine the size of the future school-age population have already taken place. Much more uncertainty surrounds future enrollment rates.

Variations in Enrollment Rates

Countries vary greatly in the proportion of the age group enrolled in primary, secondary, or higher education. There are also marked differences in enrollment rates for males and females, for urban and rural areas, and for different income groups (see table 5-3).

Figure 5-2. *Influences on Fertility*

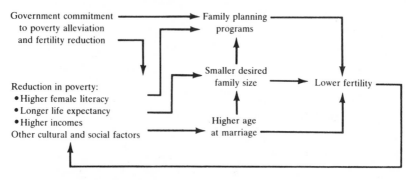

Source: World Bank (1980a), p. 66.

Although the ratio of enrollments in developing countries increased steadily during the 1970s (see table 5-4), the number of children who are not enrolled in school has also increased (table 5-5). The problem is particularly severe in the poorest countries. In some cases, school enrollment is low because facilities are not available; in other cases, young people or their families do not choose to take advantage of existing opportunities. Projections of future enrollment ratios must take into account both the planned rate of increase in the supply of school places and the effective demand for these places.

The planned rate of increase in school places has not been achieved in many countries because of financial constraints. Despite political commitment to universal primary education (UPE), rising costs and the rapid increase in the primary-school-age population have delayed the achievement of that goal in many low-income countries. A recent study of UPE in Africa (Lee 1984) has suggested that in many countries the goal could remain unrealized even by 2020 unless recurrent costs are reduced and a greater share of GNP is devoted to education. By one estimate, it would take 3 to 5 percent of GNP to achieve UPE in low-income countries (Meerman 1980).

Even when the supply of school places is sufficient to provide the opportunity for universal primary education, however, shortfalls may occur since a wide variety of other factors also affect enrollments—for example, the geographical distribution of school places, the private costs of schooling, and wastage and dropout. Added to these are socioeconomic and cultural factors (which may help to explain differences between male and female enrollments or differences between social or ethnic groups) and admissions policies (for example, restrictions on the age of entry to different levels, promotion and repetition practices, the setting of entry requirements for secondary or higher education, and the examination system).

Table 5-3. *Primary School Enrollment, by Income Group*
(percent)

Country	Boys (aged 5–9)		Girls (aged 5–9)	
	Poorest households	Richest households	Poorest households	Richest households
Sri Lanka, 1969–70	70.3	89.8	65.8	81.9
Nepal, 1973–74				
11 towns	29.5	77.8	15.3	71.2
India: Gujarat state, 1972–73				
Rural	22.7	53.9	8.6	50.9
Urban	42.1	77.7	30.8	69.5
India: Maharashtra state, 1972–73				
Rural	24.6	54.6	16.6	52.9
Urban	40.4	86.3	42.1	87.0
	Both sexes (aged 6–11)			
Colombia, 1974				
Large cities	69.6	94.6		
All urban	62.0	89.5		
Rural	51.2	60.0		

Note: Enrollments are expressed as a percentage of the number in the age group. Poorest and richest refer (in the case of India, Nepal, and Sri Lanka) to the bottom and top 10 percent of households ranked by expenditure per person, and (in the case of Colombia) to the top and bottom 20 percent of households ranked by income per person.
Source: World Bank (1980a), p. 47.

Table 5-4. *Enrollment Ratios in Primary, Secondary, and Higher Education in the Early and Late 1970s, by Region*
(percentage of age group enrolled)

Region	Primary		Secondary		Higher	
	Early	Late	Early	Late	Early	Late
East Africa	61	73	7	14	0.47	0.67
West Africa	43	43	7	9	0.36	0.73
East Asia and Pacific	94	92	28	39	2.54	6.0
South Asia	46	67	18	17	1.61	2.58
Europe, Middle East, and North Africa	74	81	25	34	2.10	4.00
Latin America and the Caribbean	93	93	26	38	7.90	6.00
All developing countries	68	75	18	25	2.50	3.30
OECD	98	100	80	80	14.6	21.0

Source: Zymelman (1982), p. 49.

Table 5-5. *Number of Children between the Ages of 6 and 11 Enrolled and out of School, by Country Income Group, 1960–75*

Country income group	1960	1965	1970	1975
Low				
Population	121.2	143.3	166.9	187.2
Number enrolled	50.9	71.3	86.6	104.7
Number out of school	70.3	72.0	80.3	82.5
Lower middle				
Population	29.3	34.0	39.2	45.0
Number enrolled	13.9	18.0	21.4	26.0
Number out of school	15.4	16.0	17.8	19.0
Intermediate middle				
Population	42.5	50.5	58.0	64.7
Number enrolled	22.5	31.7	40.5	48.2
Number out of school	20.0	18.8	17.5	16.5
Upper middle				
Population	12.7	14.3	15.7	16.9
Number enrolled	9.1	10.9	12.6	14.4
Number out of school	3.6	3.4	3.1	2.5

Source: World Bank (1980b), p. 106.

Factors Determining Private Demand for Education

Economic analysis of the private demand for education must take into account a number of factors that help to determine demand, such as the private costs of education including both earnings forgone and fees and other direct costs such as expenditure on books or materials. Also important are sex, region, the expected private benefits (in the form of increased lifetime earnings), the level of personal disposable income, and unemployment rates.

One of the most powerful influences on demand for secondary and higher education, and even on primary school enrollment rates in some developing countries, is the level of family income. Poor families will certainly find it difficult to pay fees, but even free education imposes a substantial financial burden through earnings forgone and out-of-pocket expenses for clothes, travel, books, or materials. Moreover, poor families on the average tend to have more school-age children than higher-income families. In rural areas, where many of the poorest families live, communications are likely to be difficult and there may be no access to a local school.

A study in Malaysia (Meerman 1979) concluded that effective demand at each educational level is a positive function of income. One reason is that out-of-pocket expenses, even for primary education where fees are low, represent a substantial financial burden for poor families. In 1974, the lowest income group in Malaysia on the average spent 18 percent of the total annual household income on out-of-pocket expenses for education, whereas the top income group spent less than 6 percent of family income. These figures do not take into account earnings forgone, which also represent a greater burden for low-income families. In addition, the poorest families had 2.75 children of primary or secondary school age, compared with 1.25 in the top income group. It is clear that if poor families in Malaysia choose to send their children to primary or secondary school, they must make considerable financial sacrifices. Similarly, many poor parents in India claim they do not send their children to school because they cannot afford to buy school uniforms and notebooks, and so must keep their children at home.

An equally powerful reason for keeping them at home is that poor families need the additional income that even very young children may generate. From the time they are five to six years old, children of both sexes can make important contributions to the household through housework and child care as well as productive work (Safilios-Rothschild 1980). In the Philippines and rural Bangladesh, for example, children in poor households start contributing to family income or home production at a much lower age than children in higher-income households. This translates into less chance for poor children to attend school. It has also been suggested that the economic efficiency of households in peasant societies increases with greater total work input from children (Nag 1977). This reinforces the conclusion that the value of earnings forgone, or unpaid work in the household, accounts in large part for the lack of demand for education among the poor.

In many developing countries, girls are expected to contribute to child care or home production at a much earlier age than boys. This is but one reason why girls are less likely to be enrolled in education. Many poor families regard the education of girls as a low priority, whereas the education of sons is considered an investment in security for old age. The dowry system also helps to explain differences in male-female participation. One study in Nepal, for example, found that the more daughters in a family, the more schooling the father wanted for his sons in the hope that they could contribute to their sisters' dowries (Jamison and Lockheed forthcoming).

There is some evidence that poverty and reluctance to bear the costs of educating girls reinforce each other as determinants of demand. A study in rural India (Joshi and Rao 1964) has shown that girls' participation in schooling may be much more affected by parental wealth than boys' enrollment rates. According to a recent study of private expenditure on schooling in Tanzania (Tan 1985), the direct costs of schooling—such as books, uniform, and incidental expenses—can also be higher for girls than for boys (see table 5-6).

These differences are not purely economic in origin. They have deep cultural roots. In the Middle East and North Africa, for example, religious and sociocultural traditions, such as the early age of marriage and childbearing and the unwillingness to allow girls to travel, help to explain low female participation. A principal factor, however, is the employment possibilities for girls. Where these are limited by tradition, the demand for girls' education is low, but an increase in employment opportunities for women will lead to an increase in demand for education, as was demonstrated in an experimental project in Nepal. Between 1968 and 1973 the employment of rural women as primary school teachers in Nepal increased the average enrollment of girls in primary schools from 13 percent to 25 percent (Unesco 1975).

The low rates of enrollment of girls and children from the poorest families are aspects of social selection that have been extensively analyzed by sociologists in developed countries, but until the recent work of Anderson (1983), comparative data for developing countries have been scarce.

Social Selection in the Demand for Education

Anderson investigated the influence of social selection on enrollments in secondary and higher education in approximately one-third of all countries. The principal factors he considered were the effects of the fathers' education, occupation, and income levels as indicated by published or unpublished data collected within the last thirty or forty years. The purpose of this study was to estimate the importance of social factors such as family background in determining access to schooling.

Anderson used various means to measure selectivity, such as the degree of over- or underrepresentation of different social groups; an index of dissimilarity, which measures absolute differences between the educational or occupational distribution of students' families and the distribution in the population as a whole; and a selectivity index based on percentage rather than on absolute differences. Each of these measures is concerned with a different aspect of selectivity.

Many problems arise, however, both in selecting appropriate measures, and in applying the same measures to different countries. What, for example, is the best measure of social background? The three main variables used in Anderson's analysis are father's educational level, father's occupation, and family income; these are interrelated, but by no means identical, and the degree of correlation between the variables

Table 5-6. *Mean Annual Private Expenditure on Secondary Schooling, by Item and Sex, Tanzania, 1981*
(shillings)

Item	State schools			Private schools			Ratio private/state
	Males	Females	All	Males	Females	All	
Fees	0	0	0	1,880	2,146	1,985	—
School books	126	142	131	202	240	217	1.66
Writing supplies	101	129	109	140	190	159	1.46
School equipment	93	138	107	147	188	162	1.51
Transport to school	132	193	151	266	253	261	1.73
School uniform[a]	260	280	267	272	316	289	1.08
Bedding[a]	186	248	204	252	317	278	1.36
Incidentals[b]	147	227	171	223	293	250	1.46
Total							
Excluding fees	1,045	1,357	1,140	1,502	1,797	1,616	1.42
Including fees	1,045	1,357	1,140	3,382	3,943	3,601	3.16

—Not applicable.
a. Includes laundering costs.
b. Includes cost of toothpaste, soap, and sundry items.
Source: Tan (1985).

varies between countries. Equally important, the relationship between the three variables and social status also varies between countries, particularly between countries at different stages of economic development. Quite apart from the problem of ensuring consistent definitions, there is the chance that an occupation such as farmer or manual worker may have a high status in one country and a low status in another.

This means that it is difficult to analyze the effect of rising per capita income on social selectivity, measured by a father's occupation. In the least developed countries, the category of manual worker commonly refers to wage workers in urban settings who are privileged when compared with landless peasants. The status advantage is not maintained at later stages of economic development, however, even though the category of manual worker may become increasingly skilled.

Considerable differences also exist in the status and the relative incomes of farmers in countries at different stages of development. One of the findings of Anderson's study was that in higher-income countries, where agriculture employs only a small fraction of the working population but yields an average or above-average income, the selectivity index for children of farmers exceeds that for children of manual workers. By contrast, in poorer countries, even unskilled manual workers in the modern sector are part of a privileged elite, and their children are likely to enjoy far greater access to schooling than children of landless agricultural workers.

Social selectivity is also influenced by factors not examined in Anderson's study, such as ethnic group, religion, and language, which may be important determinants of educational opportunities. Social selectivity is, however, only loosely correlated with measures of economic development such as per capita income. At each stage of economic development, the degree of social selectivity varies greatly between countries with similar levels of per capita income. Time-series data from some countries indicate a trend toward greater equality of access for different social groups. Increases in overall enrollment rates at the secondary level have played a large part in the democratization, but changes in the structure of occupations and the educational distribution of the adult population have also played a role.

Social disparities have also been reduced at the university level, but not as much as in secondary schooling, except in a few cases (for example, Greece and Puerto Rico) where tremendous progress toward equalization has taken place. Attempts to improve equality of opportunity at lower levels may increase the degree of selectivity at higher levels of education, however. A study in Nigeria concluded that the introduction of free primary education, which was intended to equalize educational opportunities, had the effect of increasing competition at the secondary level (Abernathy 1969). In other words, commitment to equality at one level made it more difficult to realize equality at the next level. Abernathy observed that as inequalities in selection for secondary school became more apparent, Nigerians "tended to ignore the egalitarian impact of universal primary education and to concentrate instead on the ways in which the educational system was stratifying a traditionally classless society" (p. 252).

Investment decisions may thus have unanticipated consequences for social equality and for equity (see chapter 9). Social origins are a powerful determinant of the private demand for education, and they are often linked with other important determinants of demand, such as attitudes and values. For example, inequalities of participation in education may reflect the different values ascribed to education by different social groups. This brings us back to the question of private costs and benefits, and the way these are perceived by students or their parents. As research has shown, parents' perceptions of the costs and benefits of education for girls and boys may account for the observed sex differentials in enrollment rates. It is difficult to obtain data on *perceived* costs and benefits and *expected* rates of return, as opposed to *actual* rates of return. One recent study (Tan 1985) approached the problem by collecting data on secondary school pupils' own estimates of earnings forgone and expected future earnings and found, surprisingly, that the expected private rate of return to secondary schooling was higher among girls than boys. Although these expectations may not prove to be realistic, they do influence individual investment decisions and define what is sometimes called taste for schooling.

Anderson (1983) examined the significance of "educogenic" families in this regard; these families have a strong aspiration for schooling for their children because older members of the family have been well schooled, whereas in other families this taste for schooling is absent. He found that the level of parental schooling is only one of many factors that may influence taste for schooling. Some others are geographic location and proximity to schools, ethnic group, kinship patterns and family size, and language. Ethnic group is extremely important in some developing countries. In Ghana, for example, the enrollment of boys in secondary school differs considerably among tribes (Foster 1965), while in Malaysia there are marked differences among ethnic groups. The government of Malaysia has introduced a wide variety of measures, including quotas, to reduce disparities between the Malays and Chinese.

Another question that arises in a multilingual society is whether primary schooling should be conducted in a child's first or second language. A recent review of the research in this area (Dutcher 1982) notes some conflicting results, but finds that overall the language of instruction may be a determinant of a child's achievement in primary school and therefore may subsequently determine demand for secondary or higher education. Although there is no "best answer" to the question of which language to choose, factors that must be taken into account include the linguistic and cognitive development of children in their first language, the attitudes of parents toward the language chosen for the school, and the status of the languages in the wider community.

Economic factors, of course, also have a great effect on the private demand for education. Among these are job opportunities and individual perceptions of costs and benefits, or private rates of return. We turn now to the question of how to estimate and interpret the private rate of return, and how the results help to guide individual investment decisions.

The Private Rate of Return to Investment in Education

Just as the social rate of return provides a convenient summary statistic showing the relation between costs and benefits of educational investment for society as a whole (see chapter 3), so the private rate of return measures the relation between costs and economic benefits for the individual. Private costs include both the direct and indirect costs of education for the individual pupil, student, or family. The direct costs include fees and expenditure on books, materials, school uniform, travel to and from school, and other out-of-pocket expenses. The importance of fees varies considerably among countries, but even when tuition is free, out-of-pocket expenses on books, clothes, and other similar items may be substantial. In some developing countries, such as Mauritius, primary and secondary schooling is free, but parents still incur considerable private expenditures on private tuition to supplement regular school instruction. In addition to the direct costs of education, the individual student or his family must also bear the indirect cost of earnings forgone, or the loss of a student's productive work while in full-time education. Earnings forgone can be derived from age-earnings profiles, and are usually measured by the average earnings of those with lower levels of education; thus the earnings forgone of university students are measured by the average earnings of secondary school leavers.

In many countries, the private costs of education may be reduced through financial support to students in the form of grants, bursaries, scholarships, or loans. The average level of financial aid is therefore subtracted from average private expenditure to give the average private costs of education. This figure can then be compared with expected private benefits to show the profitability of education as an investment for the individual.

Private benefits are measured by the additional lifetime earnings of educated workers and are derived from age-earnings profiles in the same way as social benefits. The only difference is that private benefits (and earnings forgone) are measured after payment of taxes, whereas social benefits are measured in terms of gross, pretax earnings. Other adjustments may be made to age-earnings profiles. Factors that should be taken into account, for example, are the alpha coefficient, which estimates the proportion of earnings differentials due to education (rather than ability or family background), and the probability of graduates or school leavers being unemployed.

The question of *why* educated workers earn more than the uneducated is less crucial in calculating the private benefits of education, however, than it is in estimating social returns. The assumption that the higher earnings of educated workers are due to their higher productivity is crucial to social cost-benefit analysis, but in the analysis of the private rate of return this assumption is not necessary. Provided that educated workers receive higher pay than the uneducated, education is profitable for the individual and offers financial benefits even if employers are completely irrational in choosing to pay higher salaries to the educated or even if the screening hypothesis (see chapter 3) accounted for the entire income differential of educated workers.

Since the evidence on private rates of return has already been presented (see table 3-11), the main conclusions can simply be reiterated here:

- The private rate of return is consistently higher than the social rate of return.
- The private returns to primary education are by and large well in excess of 15 percent, and may be as high as 50 percent. In the case of secondary and higher education, estimates of private rates of return are also high, usually well in excess of 10 or 12 percent, and often as high as 30 or 40 percent.

Thus, for the individual student or family, education is usually a highly profitable personal investment. The expected benefits more than compensate for the burden of high costs, including earnings forgone, and the fact that the social rate of return is always lower than the private rate of return indicates that education is highly subsidized, and that the extra taxes paid by the educated do not compensate for this subsidy.

If individuals act rationally and choose between alternative investment opportunities on the basis of private rates of return, then there is likely to be strong demand for education, particularly at the primary level, where private rates of return are usually extremely high. But this is also likely to be the case at the secondary or university level, where rates of return frequently exceed the probable returns on alternative investment opportunities. The fact that private rates of return are often as high as 20 to 30 percent or more means that it would be profitable for individual students or their families to borrow in order to pay the private costs of education, provided that the cost of borrowing is lower than the private rate of return. In many developed and developing countries where student loans are available to help individuals meet the private costs of education (see chapter 6), the evidence suggests that even if students had to pay interest rates of 10 or 12 percent on these loans, most types of education would still be profitable. In most cases, however, interest on student loans is subsidized, so that students pay rates of interest considerably below the market rates and thus obtain even greater private rates of return.

If individuals are trying to maximize private rates of return, or even if they have a rough idea of private rates of return, this will influence their decision on whether to enroll in education. But do most individuals or families regard educational decisions as investment decisions? The notion that an uneducated peasant in a developing country chooses whether or not to send his child to school on the basis of estimates of the private rate of return to education, may seem fanciful or absurd. Yet many are deeply influenced by their perception of the relative costs and benefits of education as a personal investment even though they may not make actual calculations of expected rates of return.

By examining socioeconomic and demographic aspects of school enrollment and attendance, recent research on the household economy in Botswana (Chernichovsky 1981), Tanzania (Tan 1985), and Malawi (Tan, Lee, and Mingat 1984) has thrown light on the actual and perceived private costs and benefits of schooling in developing countries. Studies of family life and rural development in Botswana have shown that, traditionally, schooling was perceived as a burden on the family, both because of the direct costs (fees or travel expenses) and the loss of the child's contribution to household production and income. At the same time, many parents failed to perceive and realize the benefits of education. The private costs and benefits have been affected by recent economic and policy changes in Botswana, however. In 1978, for example, the government abolished primary school fees and thus reduced the costs; in addition, schools now often provide free meals and health care for pupils, thus increasing the immediate private benefits of schooling.

When Chernichovsky examined how these changes had altered perceived costs and benefits and the demand for schooling, he found that enrollment rates in rural Botswana are higher among girls than boys, despite the fact that education has a greater effect on the wages of boys than girls (table 5-7). Since earnings forgone are also higher for boys than for girls, however, the higher private costs outweigh the higher expected benefits. This helps to explain seasonal variations in school enrollments (the only month when boys are more likely than girls to be enrolled in school is July, when girls traditionally help with the beginning of the harvest).

The fact that school enrollment in Botswana also tends to increase

Table 5-7. *School Enrollment, by Sex and Age Group, in Rural Botswana*

Age group	Boys	Girls	Total
7–9			
Percent	44.1	59.6	52.1
Total	261	282	543
10–14			
Percent	53.3	64.4	58.8
Total	461	472	933
15–18			
Percent	32.4	36.5	34.7
Total	241	326	567
All			
Percent	45.5	54.7	50.4
Total	963	1,018	2,043

Source: Chernichovsky (1981), p. 4.

Table 5-8. *Expected Private Rates of Return to Secondary Schooling, by Subject Area, Sex, and Type of School, Tanzania, 1981*

Subject area	Males		Females	
	State schools	Private schools	State schools	Private schools
Academic	7.6	18.3	16.1	23.1
Commercial	16.3	20.8	17.3	18.1
Technical	14.9	18.9[a]	24.8[b]	22.5[c]
Agriculture	15.1	17.1	20.7	17.8

a. Based on 56 observations.
b. Based on 25 observations.
c. Based on 5 observations only.
Source: Tan (1985).

when there are elderly people in the household suggests that the earnings forgone of children are less important for the family budget when substitutes for child labor are available. This then reduces the costs of schooling and increases demand. Holdings of cattle, land, and small stock are also correlated with school enrollment; land or stock holdings increase the opportunity cost of children's time and therefore reduce the demand for schooling, whereas large cattle holdings, which in rural Botswana are an indication of family wealth, mean that families can afford to send their children to school even though the opportunity cost of their time is high.

This situation in Botswana or elsewhere can be analyzed in terms of the conflicting effects of price and wealth on demand for schooling. In normal circumstances, an increase in price has a negative effect on demand, whereas an increase in wealth has a positive effect. In other words, the higher the price of education (including both fees and earnings forgone), the lower the demand for schooling, and the greater the level of family wealth, the higher the demand. On the other hand, in rural Botswana, where wealth is measured by holdings of land or cattle, greater wealth also creates a greater demand for child labor and thus increases the opportunity cost of attending school. In some cases the negative price effect is more powerful, but, if children from wealthier families expect a higher private rate of return to schooling than poorer children (because they expect higher earnings), this expectation will increase their demand for schooling. A survey of secondary school pupils in Tanzania (Tan 1985) has shown that private school pupils who come from upper-income families have higher earnings expectations than pupils in government schools, so that their expected private rates of return are higher, even though the private costs of secondary schooling are higher in private schools than in government schools (see table 5-8).

Estimates of expected rates of return are based on pupils' earnings expectations and actual costs. Pupils or their families do not, of course, actually sit down and calculate private rates of return. Nevertheless, students in many developing countries or their families often do have clear perceptions of private costs and benefits, and these influence demand for schooling.

These perceived costs or benefits may differ from actual costs and benefits, however, one of the reasons being that they are ex ante rather than ex post rates of return based on expected earnings. Students do not usually have reliable up-to-date information on earnings or may over- or underestimate costs or benefits for some other reason. Accurate estimates of private benefits of education should take into account not only the earnings differentials associated with education, for example, but also the differences between the educated and uneducated in the probability of employment. Potential students or their families may over- or underestimate both of these, although it is interesting that expected rates of return calculated on the basis of pupils' earnings expectations in Malawi (Tan, Lee, and Mingat 1984) were fairly close to private rates of return calculated from actual earnings data.

To find out whether students' perception of benefits is accurate and whether they are more influenced by expected earnings differentials, employment chances, or other factors, some tracer studies have examined the motivation of those who choose to enroll in higher education. Studies conducted by the International Institute for Educational Planning (IIEP), for example, found that "desire for professional qualifications" was by far the most common reason that students gave for choosing higher education in the Philippines, Sudan, Tanzania, and Zambia (Psacharopoulos and Sanyal 1981). Significant differences in motivation are evident, however, even though all students attach great importance to professional qualifications. Table 5-9 shows that in the Sudan, for example, the proportion of students giving "need for professional qualification" was higher among the children of farmers or government employees than among the children of unskilled workers.

Evidence from Tanzania indicates that the salary expectations of students in higher education are strongly influenced by starting salaries in the public sector. Students with secondary qualifications, for example, expected a salary of 625 shillings a month while those with postsecondary qualifications expected 1,600 shillings a month. The actual starting salaries of civil servants in Tanzania are remarkably close to these expected salaries.

The educational and career choices of students may therefore be influenced by their knowledge of starting salaries, rather than by their expectations of lifetime income. Even if students lack reliable data on

Table 5-9. *Reasons for Continuing to Higher Education, Sudan*
(percent)

Father's occupation or income	Need for career qualification	Lack of employment opportunity
Occupation		
Farmer	67	5
Merchant	55	7
Government employee	70	3
Skilled worker	60	0
Unskilled worker	54	16
Income		
Under 250 pounds	65	6
250–500	64	5
500–1,000	70	5
More than 1,000	47	3

Source: Psacharopoulos and Sanyal (1981), p. 48.

average lifetime earnings, they may have quite accurate information on starting salaries, which could be used to arrive at a rough comparison of costs and benefits and to assess the likely profitability of higher education as a personal investment. In many countries, public sector pay scales influence student expectations and private demand for education and their perceptions of private rates of return. Students do not actually calculate rates of return, but they often have a fairly accurate idea of the balance between costs and benefits and therefore can be said to act "as if" they were influenced by private rates of return. The private rate of return helps to explain why, for example, there is still strong private demand for education in developing countries where educated unemployment is prevalent. A study of graduate unemployment in India (Blaug, Layard, and Woodhall 1969) found that the private rates of return to secondary and higher education were still well in excess of most alternative investment opportunities even when allowances were made for the probability of unemployment. Thus, demand for school and university places is strong in India despite the relatively high incidence of graduate unemployment. This and other evidence on rates of return in India led Heyneman (1979) to conclude that it is reasonable for individuals to invest in formal schooling in India.

In fact, far from proving that education is no longer a profitable private investment, analysis of graduate unemployment in India provides further evidence of the importance of public sector pay scales, particularly starting salaries, in determining students' perceptions of the costs and benefits of higher education. The rigidity of public sector pay scales, combined with the lack of job mobility, increases graduate unemployment by encouraging graduates to wait for long periods for their first job, rather than accept a job with a low starting salary. This is perfectly rational behavior, since starting salaries are a powerful determinant of lifetime income in India. Thus, it may make perfectly good sense to wait a little longer rather than to accept a low starting salary that will determine subsequent lifetime earnings. A study of unemployment among secondary school leavers in Indonesia (Clark 1983) also shows that, as in India and in many other developing countries, educated unemployment largely consists of an extended waiting period for a first job.

This means that government may be able to exert a powerful influence on private demand for education by changing students' perceptions of the benefits of education, for example, through public sector pay scales. The fact that students are strongly influenced by starting salaries, particularly in the public sector, means that civil service pay is often a powerful tool for influencing perceived private benefits and rates of return to educational investment.

Trends in public sector pay scales such as those in Kenya—where the government substantially increased the salaries paid to graduate civil servants between 1968 and 1971—show why there has been strong and increasing private demand for education in developing countries in recent years. A calculation of rates of return based on 1968 civil service salary scales in Kenya suggested rates of return of more than 20 percent (Rogers 1972), but by 1971 they were upwards of 30 percent as a result of the revised salary scales (Fields 1974).

Many governments are now concerned that this strong private demand for education is exerting pressure on educational expenditure at a time of growing financial constraints. The conclusion from private rates of return is clear: if governments wish to reduce the pressure from private demand for education, they should act to influence both the expected private costs and benefits.

We have already suggested that a powerful lever on the benefit side is public sector pay scales, particularly starting salaries, which influence students' perceptions of the financial benefits of secondary or higher education. On the cost side, changes in the level of subsidy, through increased tuition fees or charges for meals, accommodation, or health care, will all affect the private costs of education, as will changes in student aid policy (for example, the substitution of loans for scholarships, bursaries, or grants).

The estimated private rates of return in Kenya suggest that higher education is a very profitable investment for the individual student. Thus both Fields and Rogers argue that the levels of subsidy should be reduced through the introduction of student loans, in place of the generous system of free tuition, books, room, and board. A more recent investigation (Armitage and Sabot forthcoming) has concluded that subsidies for government secondary schools in Kenya should be substantially reduced. The very strong private demand for schooling in Kenya has led to the rapid expansion of relatively unsubsidized private and harambee schools. (Harambee is a Swahili word meaning "let's pull together"; harambee schools are those built and financed by the community.) Private costs are considerably higher than in government secondary schools, but the quality, and therefore expected benefits, are lower. The result is that the private rate of return to harambee schools is markedly lower than the private rate of return to government schools. Armitage and Sabot estimate that government schools yield returns 50 percent higher than harambee schools, and therefore argue that fees could be raised substantially in government schools without reducing demand. Similarly, Tan, Lee, and Mingat argue that private rates of return, and therefore private demand, for secondary education in Malawi are sufficiently high to justify an increase in tuition fees in secondary schools. The way education is financed clearly affects students' perceptions of costs and benefits and their judgments about the relative profitability of different types of educational investment and therefore helps to determine private demand for education; the demand in turn influences government decisions on educational investment.

References

Abernathy, David. 1969. *The Political Dilemma of Popular Education: An African Case*. Stanford, Calif.: Stanford University Press.

Anderson, C. A. 1983. Social Selection in Education and Economic Development. Washington, D.C.: World Bank, Education Department.

Armitage, J., and R. H. Sabot. Forthcoming. Efficiency and Equity Implications of Subsidies of Secondary Education in Kenya. In *Modern Tax Theory for Developing Countries*, ed. David Newbery and Nicholas Stern. New York: Oxford University Press.

Blaug, M., P. R. G. Layard, and M. Woodhall. 1969. *The Causes of Graduate Unemployment in India*. London: Allen Lane, Penguin Books.

Chau, Ta Ngoc. 1969. *Demographic Aspects of Educational Planning*. Paris: Unesco/International Institute for Educational Planning.

Chernichovsky, Dov. 1981. Socioeconomic and Demographic Aspects of School Enrollment and Attendance in Rural Botswana. Population and Human Resources Division Discussion Paper no. 81-47. Washington, D.C.: World Bank.

Clark, David. 1983. *How Secondary School Graduates Perform in the Labor Market: A Study of Indonesia*. World Bank Staff Working Paper no. 615. Washington, D.C.

Cochrane, Susan H. 1979. *Fertility and Education: What Do We Really Know?* Baltimore, Md.: Johns Hopkins University Press.

Davies, David. 1980. Human Development in Sub-Sahara Africa. In *Poverty and the Development of Human Resources: Regional Perspective*, ed. W. Bussink et al. World Bank Staff Working Paper no. 406. Washington, D.C.

Dutcher, Nadine. 1982. *The Use of First and Second Languages in Primary Education: Selected Case Studies.* World Bank Staff Working Paper no. 504. Washington, D.C.

Fields, Gary S. 1974. Private Returns and Social Equity in the Financing of Higher Education. In *Education, Society and Development: New Perspectives from Kenya*, ed. D. Court and D. P. Ghai. Nairobi: Oxford University Press.

Foster, Philip. 1965. *Education and Social Change in Ghana.* London: Routledge.

Grawe, R. 1980. Human Development in South Asia. In *Poverty and the Development of Human Resources: Regional Perspectives*, ed. W. Bussink et al. World Bank Staff Working Paper no. 406. Washington, D.C.

Heyneman, Stephen P. 1979. *Investment in Indian Education: Uneconomic?* World Bank Staff Working Paper no. 327. Washington, D.C.

Jamison, Dean T., and Marlaine E. Lockheed. Forthcoming. Participation in Schooling: Determinants and Learning Outcomes in Nepal. *Economic Development and Cultural Change.*

Joshi, P. C., and M. R. Rao. 1964. Social and Economic Factors in Literacy and Education in Rural India. *Economic Weekly*, January 1964.

Lee, Kiong Hock. 1984. Universal Primary Education: An African Dilemma. Washington, D.C.: World Bank, Education Department.

Liu, B. A. 1969. *Estimating Future School Enrollment in Developing Countries: A Manual of Methodology.* New York: Unesco/United Nations, 1966.

Meerman, Jacob. 1979. *Public Expenditure in Malaysia: Who Benefits and Why?* New York: Oxford University Press.

———. 1980. Paying for Human Development. In *Implementing Programs of Human Development*, ed. P. T. Knight. World Bank Staff Working Paper no. 403. Washington, D.C.

Nag, M., et al. 1977. The Economic Value of Children in Two Peasant Societies. *Proceedings of the International Population Conference.* Liège, Belgium: International Union for the Scientific Study of Population. Quoted in Safilios-Rothschild (1980).

Psacharopoulos, George, and Bikas Sanyal. 1981. *Higher Education and Employment: The IIEP Experience in Five Less Developed Countries.* Paris: International Institute for Educational Planning.

Robbins Committee on Higher Education. 1963. *Higher Education: Report of Committee on Higher Education.* London: Her Majesty's Stationery Office.

Rogers, Daniel. 1972. Student Loan Programs and the Returns to Investment in Higher Levels of Education in Kenya. *Economic Development and Cultural Change* 20, no. 2 (January):243–59.

Safilios-Rothschild, Constantina. 1980. The Role of the Family: A Neglected Aspect of Poverty. In *Implementing Programs of Human Development*, ed. Peter Knight. World Bank Staff Working Paper no. 403. Washington, D.C.

Tan, Jee-Peng. 1985. The Private Direct Cost of Secondary Schooling in Tanzania. *International Journal of Educational Development* 5, no. 1:1–10.

Tan, Jee-Peng, Kiong Hock Lee, and Alain Mingat. 1984. *User Charges for Education: The Ability and Willingness to Pay in Malawi*. World Bank Staff Working Paper no. 661. Washington, D.C.

Unesco. 1975. *Women, Education and Equality: A Decade of Experiment*. Paris.

World Bank. 1980a. *World Development Report 1980*. New York: Oxford University Press.

———. 1980b. *Education*. Sector Policy Paper. Washington, D.C.

Zymelman, Manuel. 1979. *Patterns of Educational Expenditures*. World Bank Staff Working Paper no. 246. Washington, D.C.

———. 1982. Educational Expenditures in the 1970s. Washington, D.C. World Bank, Education Department.

6

Financing Educational Investment

Education is both a private and a social investment that is shared by individual students, their families, employers, government, and other groups, including international agencies. The sharing arrangements vary considerably from country to country, both in the proportions of public and private funds allocated to education and in the mechanisms by which the costs of education are financed. During the 1960s and 1970s, most of the expansion of education was financed by increased public expenditure on education, which rose in relation to national income and public expenditure as a whole. According to World Bank estimates, the proportion of GNP devoted to education in developing countries rose on the average from 2.3 percent in 1960 to 4.5 percent in 1984, and the proportion of the national government budget rose from 11.7 percent in 1960 to 16.1 percent in 1984.

The fact that educational investment took an increasing share of the national budget reflected the high priority given to education. Governments believed it would promote economic growth and provide the skilled manpower needed for development. Thus in the early 1970s Jallade (1973) was led to predict that education expenditure would absorb a growing share of both the budget and GNP in developing countries. Before long, however, there were warnings of an impending crisis in the financing of education. For one thing, attitudes toward education spending began to change, partly as a result of the huge increases that had occurred in the 1960s. Many also expressed uncertainty about the economic role of education in the light of diminishing scarcity of qualified manpower in developing countries and the growing concern about the brain drain and unemployment among the educated. Moreover, other sectors such as health, population, nutrition, and rural development began to challenge educational investment for public funds.

Today, there is increasing evidence of financial constraints, and in many developing countries the proportion of the government budget and GNP devoted to education has begun to decline. According to Zymelman

(1982), average government expenditure on education in Africa, East Asia, and the Pacific during the early and late 1970s remained fairly constant as a proportion of all government expenditure, whereas in South Asia, Latin America, North Africa, and the Middle East, the proportion of government expenditure devoted to education fell after 1975 (table 6-1).

Although the social and private rates of return to investment in education still seem to be high in many countries and the private demand remains strong, governments are no longer willing to allocate an increasing share of public expenditure to education. In some countries, this attitude reflects increased skepticism about the economic benefits of education and concern about unemployment among the educated. In others, higher priority is being given to other forms of social investment, such as health care or agricultural development. In some developing countries the decline in government expenditure on education is simply a result of the world recession. For a variety of reasons, then, education is facing growing financial constraints.

As a result, more effort is being put into research and analysis of alternative methods of financing education, particularly cost recovery and the redistribution of the financial burden of investing in education. The World Bank (1980) has emphasized that the increasing demands of education on public finance at a time when government funds are stagnant or even falling in many developing countries can only be resolved by either finding additional sources of financial support or reducing unit costs through greater efficiency. Developing countries are exploring both possibilities. This chapter is concerned with the first possibility—that changes in the system of financing educational investment might help to reduce the pressure on public funds (the second is discussed in chapter 8).

Table 6-1. *The Mean Share of Education Expenditures in Government Budgets, by Region*
(percent)

Region	1972–75	1976–79
East Africa	15.1	15.1
West Africa	15.6	16.1
East Asia and Pacific	18.3	18.4
South Asia	8.1	7.1
Latin America and the Caribbean	18.2	16.5
Europe, Middle East, and North Africa	14.6	13.5
All regions (developing countries)	15.9	15.2
OECD	9.6	9.0

Source: Zymelman (1982), p. 50.

First, however, we should look at the present pattern of financing education and consider why such a high proportion of the resources devoted to education comes out of public funds.

The Balance between Public and Private Financing of Education

Because there is little information about private expenditure on education, it is difficult to measure the precise contribution of private financing, which may take the form of tuition fees, registration or examination fees, and the purchase of books, materials, and special clothing (such as uniforms), as well as the indirect costs of earnings or income forgone. National account statistics sometimes provide data on private expenditure on education (see table 6-2), but in many cases either the education expenditure is included with expenditure on recreation, entertainment, and cultural services, or the estimates are questionable. Nonetheless, the data suggest that private expenditures on education vary considerably in

Table 6-2. *Private Expenditure on Education as a Percentage of Total Private Expenditure*

Region/country	1960	1965	1970	1975	1978
Africa					
Ghana[a]	n.a.	n.a.	n.a.	3.8	n.a.
Mauritius[a]	n.a.	n.a.	n.a.	5.8[b]	n.a.
Niger[a]	n.a.	1.3	n.a.	n.a.	n.a.
Sierra Leone	n.a.	3.0	3.1	n.a.	n.a.
South Africa	0.8	0.9	1.1	1.3	1.3
Sudan	n.a.	n.a.	0.8	0.8	n.a.
Tanzania[a]	n.a.	2.5	2.0	n.a.	n.a.
Togo	n.a.	0.4	0.6	n.a.	n.a.
Zambia[a]	n.a.	1.3	1.4	n.a.	n.a.
Zimbabwe	n.a.	n.a.	4.2	3.1	3.0[c]
Asia and Pacific					
Fiji	n.a.	n.a.	3.2	2.7	2.7[c]
Hong Kong	n.a.	2.3	2.0	1.8	1.3
India	2.2	2.6	3.0	3.2	3.2
Kampuchea[a]	n.a.	4.6	n.a.	n.a.	n.a.
Kiribati	n.a.	n.a.	n.a.	1.2	n.a.
Korea, Rep. of	1.6	0.8	2.8	2.4	3.0
Malaysia	1.4	0.7	0.8	n.a.	n.a.
Papua New Guinea	n.a.	0.5	0.5	n.a.	n.a.
Philippines	3.2	2.9	3.2	2.7	n.a.
Singapore	0.9	1.3	1.1	0.8	0.8
Sri Lanka	n.a.	n.a.	n.a.	1.1	1.2
Thailand[a]	4.1	7.2	8.3	8.9	9.2

Table 6-2 *(continued)*

Region/country	1960	1965	1970	1975	1978
Latin America and the Caribbean					
El Salvador	1.7	3.5	2.3	2.1	1.9c
Honduras[a]	5.2	6.1	6.0	5.1	n.a.
Jamaica	0.9	0.8	1.0	0.4	0.4c
Panama[a]	5.8	5.9	8.2	7.8	7.0c
Puerto Rico	1.0	0.8	0.9	1.0	1.2
Venezuela[a]	n.a.	n.a.	n.a.	8.9	9.2
Europe, Middle East, and North Africa					
Cyprus	n.a.	n.a.	0.9d	0.6	0.6c
Israel	4.2	4.7	6.2	6.1	6.3
Jordan	1.2	1.5	1.8	2.6	2.6c
Libya[a]	n.a.	3.5	3.7	n.a.	n.a.
Malta[a]	5.6	6.1	7.0	8.2	8.2
Selected OECD countries					
Australia	0.7	0.9	0.8	0.5	0.5c
Belgium	0.3	0.2	0.2	0.2	0.2
Canada	1.1	1.6	2.3	2.3	2.8
France	n.a.	0.6	0.3	0.3	0.3
Greece	2.2	1.6	1.6	1.8	1.6
Japan[a]	n.a.	9.3	9.0	9.0	9.0
Spain	1.3	1.7	2.1	2.2	2.3c
United Kingdom	1.3	1.8	2.1	2.2	2.3
United States	1.3	1.6	2.0	2.0	1.9

n.a. Not available.

a. Figures include expenditure on recreation, entertainment, education, and cultural services.

b. 1976.

c. 1977.

d. 1971.

Source: Tan (1985b).

developing countries, from less than 1 percent to about 3 to 4 percent of total private consumption expenditure. These differences have arisen in part because private schools in some countries charge tuition fees, whereas in others there are no tuition fees.

Estimates of enrollment in private primary and secondary schools (table 6-3) suggest that the proportion of pupils in private schools in 1965 and 1975 was much higher at the secondary level than at the primary level, but that the proportion of enrollment in private schools declined between these years. Similar data do not exist for higher education, although in some developing countries (for example, Colombia and the Philippines), private universities are important. This general decline is confirmed by two regional studies of Africa and Latin America. With only a few exceptions, private enrollments in various African countries

Table 6-3. *Percentage of Students Enrolled in Private Schools, 1965 and 1975*

Region/country	Primary level 1965	1975	Secondary level 1965	1975
Asia				
Bangladesh	n.a.	8	n.a.	9
Indonesia	12	13	n.a.	60
Korea, Rep. of	1	1	48	45
Philippines	4	5	66	38
Singapore	40	35	3	1
Sri Lanka	n.a.	6	9	n.a.
Thailand	13	11	50	32
East Africa				
Botswana	4	5	10	30
Burundi	96	92	30	22
Ethiopia	25	25	15	n.a.
Kenya	4	1	29	49
Lesotho	96	100	100	89
Madagascar	27	23	66	49
Malawi	77	10	5	13
Mauritius	34	28	77	6
Sudan	2	2	45	13
Swaziland	80	80	4	n.a.
Tanzania	7	4	n.a.	29
Zaire	91	n.a.	57	n.a.
Zambia	n.a.	24	4	2
West Africa				
Benin	40	5	54	18
Burkina Faso	34	7	38	43
Cameroon	61	43	73	57
Central African Republic	n.a.	n.a.	2	n.a.
Chad	12	10	7	6
Equatorial Guinea	n.a.	24	n.a.	3
Gabon	53	45	43	32
Gambia, The	n.a.	16	54	46
Ivory Coast	28	19	n.a.	28
Liberia	25	35	48	43
Mali	8	4	10	11
Niger	6	5	5	14
Nigeria	76	n.a.	n.a.	41
Senegal	13	12	22	n.a.
Sierra Leone	n.a.	78	n.a.	87
Togo	40	29	55	16

Table 6-3 *(continued)*

Region/country	Primary level 1965	Primary level 1975	Secondary level 1965	Secondary level 1975
Europe, Middle East, and North Africa				
Algeria	2	1	7	1
Cyprus	1	n.a.	11	13
Egypt	13	n.a.	41	22
Iran	8	8	26	17
Iraq	2	1	24	n.a.
Jordan	28	30	13	7
Libya	3	2	7	0
Morocco	6	5	14	8
Saudi Arabia	6	3	4	2
Syria	10	5	37	6
Tunisia	2	1	18	6
Turkey	1	n.a.	n.a.	2
Yemen Arab Republic	n.a.	1	n.a.	3
Latin America and the Caribbean				
Argentina	14	17	41	45
Barbados	n.a.	9	26	21
Bolivia	26	9	26	24
Brazil	11	13	49	25
Chile	27	18	38	23
Colombia	14	15	58	38
Costa Rica	4	4	24	6
Dominican Republic	7	12	n.a.	n.a.
Ecuador	18	17	38	30
El Salvador	4	6	47	47
Guatemala	19	14	54	43
Haiti	26	42	43	76
Honduras	7	5	53	51
Jamaica	n.a.	5	n.a.	9
Mexico	9	6	29	25
Nicaragua	16	15	44	n.a.
Panama	5	5	17	14
Paraguay	10	13	51	37
Peru	14	13	24	17
Suriname	n.a.	65	57	52
Trinidad and Tobago	n.a.	n.a.	41	n.a.
Uruguay	18	17	17	n.a.
Venezuela	13	11	23	18

n.a. Not available.
Source: World Bank (1980), pp. 125–26.

Table 6-4. *Total Enrollment in Private Schools in East Africa, 1965–79*
(percent)

Country	Primary level				Secondary level		
	1965	1970	1975	1979	1965	1970	1975
(Mean)	(53)	(43)	(43)	(42)	(36)	(51)	(45)
Botswana	4	5	5	2	10	59	30
Burundi	96	94	92	100	30	36	22
Djibouti	n.a.	23	13	9	n.a.	n.a.	n.a.
Ethiopia	25	28	25	18	15	n.a.	n.a.
Kenya	4	n.a.	1	n.a.	29	42	49
Lesotho	96	100	100	100	100	89	89
Madagascar	27-	20	23	n.a.	66	70	49
Malawi	77	11	10	10	5	13	13
Mauritius	34	29	28	25	77	n.a.	6
Rwanda	n.a.	n.a.	n.a.	n.a.	n.a.	n.a.	21
Seychelles	n.a.	91	8	3	n.a.	18	4
Sudan	2	4	2	n.a.	45	n.a.	13
Swaziland	80	76	80	80	4	n.a.	n.a.
Tanzania	7	2	4	0.4	n.a.	24	29
Zaire	91	n.a.	n.a.	n.a.	57	n.a.	n.a.
Zambia	n.a.	27	24	n.a.	4	n.a.	2
Zimbabwe	n.a.	n.a.	n.a.	83	n.a.	n.a.	n.a.

n.a. Not available.
Source: Tan (1985b).

fell between 1970 and 1979 (table 6-4). During the same period, private enrollments in much of Latin America lagged behind those in government schools (Brodersohn 1978).

Although some private schools derive most of their income from tuition fees, this pattern is by no means universal. In a number of countries, government subsidies are an important source of income for private institutions. According to a 1975 survey, government subsidies in Argentina ranged from 45 to 92 percent of total cost per pupil in private primary schools and between 31 and 96 percent in private secondary schools (IDB 1978, p. 158). In Ecuador, however, fees in private schools provided 87 percent of total income, other private contributions 10 percent, and government subsidies only 3 percent. Although countries differ considerably in the extent of private education and in the importance of fees, overall, the proportion of educational income from fees has declined.

Other sources of private support for educational investment are donations or endowments. These are important in some developed countries,
but are not likely to be significant in developing countries. In Bolivia, for
example, which is the only Latin American country with data on private
endowments, this source provides 11 percent of the income of private
schools (IDB 1978).

Another private source of educational support is direct labor. Local
communities may undertake to build a school, for example, or to provide
goods or services in kind (such as food or accommodation for teachers). It
is difficult to measure the value of such contributions, but so far, such
indirect financing represents only a small part of the total costs of education in most developing countries. Some interesting efforts have been
made to exploit such local resources, however, or to have pupils contribute toward the financing of their schools through direct labor, for example, by producing goods for sale.

An important source, particularly for capital investment, is external
aid. This includes World Bank loans and credit, aid from bilateral and
international agencies, and other forms of external assistance. External
aid for education increased substantially in the 1970s, and the World
Bank provided an increasing share. Despite this increase, external aid to
education on the average amounts to less than 10 percent of the total
education budget of developing countries, although the proportion of
capital investment in buildings and equipment that is financed through
external aid is often higher. Estimates of the proportion of physical
investment in Latin America financed by means of education loans from
international agencies, including the World Bank, can be seen in table
6-5. Between 1962 and 1977, the average in Latin America was only 13
percent, but external financing may have been decisive in individual
countries, especially in determining the balance between levels. Since
external aid is often in the form of loans, however, it will increase the
demand for public funds in the future as repayments become due and as
capital investments financed by means of external aid give rise to increased recurrent costs. Recipients sometimes argue that aid can be too
expensive because it generates high local costs (see World Bank 1980).
Thus in some cases external aid may eventually increase, rather than
diminish, the pressure on public funds for education and postpone, rather
than reduce, budgetary pressures.

Although private contributions and external aid are important, particularly in some developing countries, most educational support comes
from a wide variety of public (either central or local government)
sources, which range from general taxation of individuals or companies
(including taxes on income, wealth, land, property, profits, expenditure,

Table 6-5. *Education Loans of International Lending Agencies, Latin America, 1962–77*
(millions of dollars)

	Education loans at current prices					Education loans at 1976 prices (F)	Estimated public investment in education at 1976 prices (G)	Education loans as percentage of investment in education (F/G)
Year	IDB (A)	IBRD (B)	IDA (C)	AID (D)	Total (A + B + C + D = E)			
1962	12.1	—	—	0.6	12.7	24.5	379.5	6.46
1963	4.1	—	—	3.1	7.2	13.9	406.9	3.42
1964	13.1	—	—	2.4	15.5	29.9	451.6	6.62
1965	7.2	2.8	—	12.0	22.0	41.7	509.4	8.19
1966	29.1	9.5	—	1.5	40.1	73.5	549.5	13.38
1967	35.9	—	—	18.8	54.7	99.7	591.8	16.85
1968	9.4	27.3	5.1	58.8	100.6	178.7	651.7	27.42
1969	26.4	7.8	2.9	19.5	56.6	97.3	719.3	13.53
1970	13.2	15.0	—	81.9	110.1	182.4	836.7	21.80
1971	67.3	21.9	4.0	22.0	105.2	168.2	919.7	18.29
1972	28.7	9.3	5.1	13.2	56.3	88.4	1,008.6	8.76
1973	96.1	51.4	—	47.6	195.1	282.3	1,121.6	25.16
1974	19.0	43.5	3.0	20.3	85.8	101.4	1,204.0	8.42
1975	70.8	20.0	4.0	32.4	127.3	133.2	1,272.1	10.47
1976	79.4	37.5	9.5	—	126.4	126.4	1,364.1	9.27
1977	84.7	47.0	—	19.8	151.5	142.7	1,459.3	9.77

— Blank cells were not identified in the source.

Note: IDB, Inter-American Development Bank; IBRD, International Bank for Reconstruction and Development; IDA, International Development Association; AID, U.S. Agency for International Development.

Source: Inter-American Development Bank (1978), p.58.

or sales), customs and excise duties, fees and licenses, specific taxes earmarked for education (for example, property or payroll taxes, or levies earmarked for education or vocational training), to national lotteries.

Zymelman (1973) notes, however, that educational financing is still an area of experimentation, although few entirely new methods of financing have been developed. A new method that may seem to be an innovation in one country has often been used elsewhere to finance education or other types of public expenditure. National lotteries, for example, are virtually unknown in some countries, but widely used in others.

Despite differences in the way government revenue is raised and distributed, the final outcome is the same in most developing countries: by and large, expenditure for educational investment comes from public funds. Individuals or their families bear the indirect cost of earnings forgone, and usually provide funds for books and clothes, although they may be subsidized from public funds through grants, scholarships, or loans. In most developing countries the direct costs of education are financed largely by the taxpayer rather than the individual student, although tuition fees are substantial in certain cases. At this point, it seems appropriate to ask why so many governments have chosen to subsidize education so heavily.

The Arguments for Public Subsidy of Education

Three main arguments are used to justify public subsidy of education. The first point has to do with externalities (see chapter 3). That is to say, since the social benefits of education exceed private benefits, governments subsidize education to prevent underinvestment.

The second point concerns equity and equality of opportunity. If education was provided under market conditions, only those who could afford to pay tuition fees could enroll. Not only would there be underinvestment from the social point of view, but income inequalities would be preserved from one generation to the next since education is itself a determinant of lifetime income. If all individuals had access to private capital markets, then those who could not afford to pay tuition fees could borrow; if the private rate of return to investment in education was higher than the cost of borrowing, it would still be a profitable private investment. Many imperfections can be found in capital markets, however. Individual students cannot normally borrow to finance their education without providing collateral, and investment in education is risky and uncertain. Therefore governments in many countries provide loans or loan guarantees to help students finance their education. (The feasibility of student loans in developing countries is examined later in the chapter.) Governments believe, however, that externalities and equity demand substantial subsidies rather than simply the provision of loans.

Third, many also believe that education is subject to economies of scale and thus that it is more efficient to finance and provide education publicly. Although there is evidence of economies of scale (see chapter 7), differences in unit costs of education do not seem to be clearly related to its financial sources. One study, for example (Jallade 1973), found instances of high- and low-cost institutions in both public and private education.

The three arguments above support government subsidies on grounds of both efficiency and equity. They do not, however, suggest that all or even most of the costs of education must be publicly financed. What is at issue is not *whether* education should be subsidized, but to what extent. In other words, is the present balance between public and private financing optimal?

Several studies of educational financing have tried to tackle this question, but could not agree on the optimal level of public subsidy in developed or developing countries. Early studies by Rogers (1970) and Jallade (1973) analyzed the effects of alternative methods of financing and alternative combinations of public and private finance, particularly with respect to the demand for education, the total level of resources for educational investment, equity and equality of opportunity, income distribution, the internal efficiency of education, and the quality of education.

Instead of recommending a single optimal mixture of public and private financing, both studies argued that developing countries should step up experimentation with and analysis of different combinations of tuition fees, loans, and subsidies. A number of studies have subsequently been carried out on the effects of student loans in Colombia (Jallade 1974b), the feasibility of student loans in developing countries (Woodhall 1983), and the impact of tuition fees on demand for education (Birdsall et al. 1983; Mingat and Tan, forthcoming; Tan, Lee, and Mingat 1984). At the same time, some research, particularly in the United States, has focused on the equity and efficiency of different methods of finance. The purpose of all this investigation has been to throw light on three related questions: Who benefits? Who pays? Who should pay?

No simple answers are to be found, however, and educational financing remains a highly controversial subject. Several recent surveys have summarized the theoretical issues and empirical evidence on alternative systems. Bowman, Millot, and Schiefelbein (1984), for example, consider both distributive and efficiency dimensions of public support for education in Chile, France, and Malaysia.

Education can be regarded as a government investment as well as a private and social investment since governments subsidize education and in return derive higher tax revenues from the educated. To some extent, government subsidy is reflected in the difference between private and social rates of return, but this difference does not fully reflect the relationship between government costs and expected fiscal benefits. Some government expenditure—for example, student aid or tax concessions to students' parents (sometimes called fiscal exoneration)—are transfer payments, so do not appear on the cost side of a social cost-benefit analysis. If education subsidies are viewed strictly as a form of *government* investment, however, such expenditure or revenue forgone should be included. This leads to a new measure for the rate of return: the *fiscal* rate of return, which views investment from the point of view of the government, rather than society as a whole. Table 6-6 illustrates the

Table 6-6. *Costs and Benefits of Educational Investment:*
Fiscal, Social, and Private Accounts

Item	Fiscal account	Social account	Private account
Direct public outlays on higher education	Cost	Cost	—
Student support	Cost	Transfer	Offset to part (or all) of forgone earnings
Forgone earnings	—	Cost	Cost
Income tax on forgone earnings	Cost	—	Partial offset on forgone earnings
Taxes on forgone outputs	Cost	—	—
Incremental earnings attributable to higher education	—	—	Gross benefits
Incremental product attributable to higher education	—	Benefit	—
Incremental income tax due to incremental earnings attributable to higher education	Benefit	—	Deduction from gross benefits
Other taxes drawing on incremental product attributable to higher education	Benefit	—	—

— Not applicable.
Source: Bowman, Millot, and Schiefelbein (1984).

difference between the social, private, and fiscal rate of return and shows whether items represent costs, benefits, or transfers (according to whether the accounting is from the point of view of society, the private individual, or the government's tax revenue).

A few empirical estimates of the fiscal rate of return to educational investment are available, notably with respect to higher education in the United States (Eicher 1984) and in France and Chile (Bowman, Millot and Schiefelbein 1984). The estimates for France and Chile suggest that for most types of higher education the discounted future tax receipts are less than the government subsidies discounted at a rate of 10 percent. In Chile, however, discounted tax receipts are higher than subsidies discounted at a rate of 5 percent, and in France they are higher for certain types of higher education, though not for all. This means that when higher education is viewed strictly as a government investment in future fiscal capacity, it is not particularly profitable. To achieve a rate of return of 10 percent, government subsidies for higher education would have to be reduced in both France and Chile.

Recent research and analysis of the experience of developing countries in financing education falls into two main categories. During the 1970s interest was focused on questions of equity and the effects of alternative methods of financing on the distribution of costs and benefits. Because of growing budgetary pressures, interest has recently shifted to cost recovery and ways of changing the balance between public and private finance.

The Effects of Public Subsidies on Equity and Efficiency

The distributional effects of public subsidies for education have been investigated in a number of countries. Such research began in the United States with a study in California (Hansen and Weisbrod 1969), which concluded that the net effect of taxes and subsidies for higher education has been to transfer income from poor taxpayers to the rich, who are more likely to benefit from higher education. The same effect has been reported for some other states, but it has also been challenged by Pechman (1970), who reanalyzed the data for California and argued that the rich do not receive net income transfers from higher education subsidies, but on the contrary subsidize the education of middle- and low-income families.

The conclusion of Hansen and Weisbrod that education subsidies benefit the rich rather than the poor has been widely quoted and applied to other countries as proof that existing patterns of financing education are inequitable. Jallade (1973), Fields (1975), and Psacharopoulos

(1977), for example, argue that public subsidies for higher education in developing countries have the perverse effect of transferring income from poor taxpayers to rich families, whose children benefit from subsidized education.

This controversy about the distributional effects of education subsidies continues to rage in many countries and was summarized by Bowman, Millot, and Schiefelbein (1984) and Blaug (1982). One reason for the discrepancies in different studies is that results depend on how the recipients of subsidies are classified or ranked. In Santiago, Chile, for example, at least four ways have been used, and, as can be seen from table 6-7, estimates of the proportion of subsidies received by different groups depend critically on which ranking is used. Similar differences have been observed in other countries. Ideally, of course, analysis should be concerned with the effects of both taxes and subsidies on lifetime, rather than cross-sectional earnings and income, but the necessary data are not usually available.

The point at issue, as noted earlier, is who pays the taxes that finance education, who benefits from education subsidies, and would alternative methods of financing education alter the balance between the costs and the benefits of public subsidies. The problem in developing countries is not only the lack of satisfactory studies on the incidence of taxes, but also—since education subsidies are usually financed out of general taxation—the impossibility of measuring how much each income group con-

Table 6-7. *Distribution of Higher Education Subsidies in Santiago, Chile, by Classification and Ranking of Recipients*

Ranking of recipients	Percentage of aggregate subsidy accruing to different income groups		
	Lowest 25 percent	Lowest 50 percent	Highest 25 percent
Santiago households ranked in terms of schooling of head	7.0	20.0	56.0
Santiago youths ranked in terms of schooling of head of household	4.5	15.5	62.0
Santiago households ranked in terms of household income	2.0	11.5	63.5
Santiago youths ranked in terms of household income	5.5	17.0	59.0

Source: Bowman, Millot, and Schiefelbein (1984).

tributes to the financing of education rather than other types of government expenditure. Different assumptions concerning this question give rise to the disagreement between Hansen and Weisbrod (1969) and Pechman (1970), for example, on the distribution of taxes and benefits in California.

Another problem is that education subsidies involve intergeneration transfers of income as well as transfers between income groups. This is why it is desirable to use life-cycle rather than cross-sectional income data (but such data are not available for developing countries). It also explains why the assumptions that are made have a crucial effect on conclusions about the distributional impact of subsidies. Although it is impossible to be precise about the effects of public subsidies on the balance between private costs and benefits of educational investment, recent research has helped to clarify the issues and, indeed, has demonstrated that there are no clear-cut answers.

One of the first studies to analyze these questions in detail for a developing country (Jallade 1974a) examined the pattern of financing both public and private education, the incidence of taxation, and the distribution of public subsidies for education in Colombia. Private educational expenditure and enrollments are extremely important in Colombia, but private enrollments are not equally distributed between income groups, and private institutions receive some public subsidies.

The equity implications depend on whether the taxes that are used to finance public subsidies are progressive, proportional, or regressive. A tax is progressive if it takes a larger proportion of the income of the rich than of the poor taxpayers; it is regressive if the reverse is the case, and proportional if it takes the same percentage of income from all income groups. The general conclusion is that taxation as a whole is roughly proportional for most taxpayers in Colombia and progressive for only the top income groups.

The next stage in the analysis was to allocate educational subsidies between income groups and to compute subsidies received by each income group as a proportion of taxes paid. Thus it was found that in total, education subsidies in Colombia redistribute income from rich to poor since the poorest families receive more in education subsidies than they pay in taxes, whereas the richest receive only 2 percent of their taxes in the form of education subsidies (table 6-8). Subsidies for primary education in Colombia strongly benefit the poor, but those who benefit most from secondary and higher education subsidies are middle-income taxpayers. Both Jallade's analysis and another study (Selowsky 1979) suggest that the redistributive effects of primary subsidies are partly, but not wholly, offset by the negative effects of subsidies for secondary and higher education. The net result is that *households* receive more or less

Table 6-8. *Allocation of Taxes and Public Subsidies for Education among Income Groups, Colombia, 1970*

Income bracket (pesos a year)	Number of house-holds (percent)	Allocation of taxes (millions of pesos)	Public subsidies for education (millions of pesos)	Subsidies as proportion of taxes (percent)
0–6,000	19.0	223	262	117
6,000–12,000	20.2	510	424	83
12,000–24,000	24.9	1,468	1,054	72
24,000–60,000	22.9	3,108	1,717	55
60,000–120,000	8.8	2,878	672	23
120,000–240,000	3.4	2,484	252	10
Over 240,000	0.8	2,932	72	2
Total	100.0	13,603	4,453	33

Source: Jallade (1974a), p. 40.

equal subsidies, regardless of income level (table 6-9). The reason is that the redistributive effect of primary-level subsidies cancels out the regressive effect of higher education subsidies. Since poor households tend to contain more children, however, the rich benefit from higher per capita subsidies, which average 486 pesos for the top income groups, and only 280 pesos for the bottom group.

Recent studies of the distributive effects of public expenditure in Malaysia (Meerman 1979) and Indonesia (Meesook 1984) argue that present levels of subsidy at postsecondary and higher levels of education benefit the wealthy, so that a policy of shifting more of the financial burden to private rather than public funds would be justified on grounds of social equity (see chapter 9) as well as economic efficiency. These examples clearly show, however, that the effect of education subsidies on the distribution of income is extremely complex and still needs further analysis. The analysis of the Colombian situation, for example, was static, whereas a dynamic analysis was needed to provide a thorough understanding of the distributive impact of education subsidies; such an analysis would look at the effect of education subsidies on the distribution of income over time as well as the present distribution of costs and benefits. Jallade argues that the existence of a private, relatively unsubsidized sector may contribute toward a more equitable distribution of subsidies in Colombia, since the rich will be more likely to enroll in private schools, and therefore public subsidies can be concentrated on the poorer households. If private schools offer better education, then private education may be the vehicle by which to achieve *present* equity in the

Table 6-9. *Education Subsidy per Household, Colombia, 1974*
(pesos)

Income quintile (poorest to richest)	Large cities			Intermediate cities			Small towns		
	Pri-mary	Sec-ondary	Total	Pri-mary	Sec-ondary	Total	Pri-mary	Sec-ondary	Total
1	1,642	772	2,414	1,459	1,163	2,622	1,317	998	2,315
2	1,435	1,182	2,617	1,166	1,590	2,756	1,288	1,154	2,442
3	873	1,089	1,962	992	1,109	2,101	1,205	842	2,047
4	861	797	1,658	613	1,252	1,865	593	1,323	1,916
5	278	368	646	235	760	995	338	1,440	1,778
Country average	784	729	1,513	769	1,146	1,915	1,062	1,095	2,157

financing of education—but it may also turn out to be socially and economically divisive in the future.

The whole question of subsidies requires analysis of the private rate of return to educational investment by type of school and income group, but such estimates are rare in developing countries. The few that are available suggest that public and private schools differ considerably in the quality of schooling and in the private rate of return. In some countries, private schools offer a higher quality than government schools, but in several developing countries, the reverse is true: the government schools are of better quality, and many private schools are overcrowded and have poorly qualified teachers. In Kenya, for example, the private rate of return is much higher in government than in harambee schools (Armitage and Sabot forthcoming). Analysis of the effects of subsidies must therefore take quality differences into account if comparisons between the private rate of return in different types of school are to be illuminating.

An important implication of research in this area is that present patterns of subsidy for education may not be achieving either efficiency or equity objectives. Studies of the effects of subsidies in Colombia, Malaysia, Kenya, and Indonesia all suggest that the methods of financing education need to be reappraised, and the balance between public and private financing adjusted. This brings us back to the question of the feasibility of cost recovery in developing countries.

The Scope for Cost Recovery in Education

The growing financial constraints on educational investment combined with continued strong private demand for education have led several governments to consider the possibility of increasing the share of financial

Urban areas			Rural area			Country average			
Pri-mary	Sec-ondary	Total	Pri-mary	Sec-ondary	Total	Pri-mary	Sec-ondary	Uni-versity	Total
1,464	969	2,433	1,103	273	1,376	1,305	598	18	1,921
1,317	1,284	2,601	858	273	1,131	1,089	776	96	1,961
993	1,039	2,032	633	391	1,024	835	751	224	1,810
715	1,070	1,785	359	469	828	589	872	489	1,950
263	565	828	237	352	589	252	555	1,257	2,064
854	925	1,779	754	352	1,106	816	718	413	1,947

Source: Selowsky (1979).

support provided by students and their families by various cost recovery measures, including tuition fees and student loans.

Because educational financing has become an urgent problem in some developing countries, the World Bank is currently engaged in exploring the feasibility and the implications of various cost recovery mechanisms. The World Bank's position on cost recovery was explained in a recent report (World Bank 1982, pp. 68–69) on the work of the International Development Association (IDA):

> If a project is genuinely productive, beneficiaries should be able to pay for the costs of goods and services it provides and still increase their real incomes. To the maximum extent feasible, therefore, IDA wants the costs of its projects to be recovered. This is not because IDA is concerned about having its loans repaid; they are repaid by the recipient government no matter what rate of return or degree of cost recovery is achieved on the project. More important issues are involved when projects fail to pay for themselves:
>
> - The recipient may not be able to maintain the project and will be unwilling to repeat it elsewhere.
> - The project beneficiaries will be receiving an income subsidy not enjoyed by others.
> - Materials and services received for free or at prices below their true cost may be wasted or used inefficiently.
>
> User charges raise controversial questions about equity and are often difficult to levy. . . . Nonetheless, if governments have the political will and administrative capacity, they can recover costs.

In a review of its experience with agricultural and rural development projects, the Bank's West African regional department commented on the apparently growing inability of some governments to finance recurrent costs and concluded that effective cost-recovery systems would be essential if the projects were to have a lasting impact on rural economies, but admitted this would require an "unpalatable political decision" on the part of the governments concerned. This conclusion on the need for cost recovery through user charges is just as relevant to education investment because some governments have displayed a growing inability to finance recurrent education costs. But it is also true that many governments regard the introduction of user charges as politically "unpalatable" or even "suicidal." Those that are looking into ways of introducing or increasing cost recovery and attracting new private sources of funds are considering several possibilities: tuition fees in public sector schools; growth of the private sector, where fees may already be charged; fees or charges for accommodation, food, or other services; payroll taxes, levies, or other financial contributions from employers to help finance vocational training or adult education; increased provision of vocational training by employers; direct labor or contributions in kind from the local community; student loans; graduate or professional surtax; and vouchers.

A recent review (Meerman 1980) of the financial requirements of human development programs (including education, health care, nutrition, water supply, and sanitation) and the likely growth of tax revenues in developing countries concluded that if low-income countries rely chiefly on central government resources, most of them will be unable to provide a minimum effective standard of human development for their population as a whole. Meerman points out that the scope for increasing general tax revenues is very limited in low-income countries; the average tax-GNP ratio in developing countries has already increased from 11.4 percent in 1953–55 to 16 percent in 1972–76, but has been sluggish since then owing to the oil crisis and world recession (Tait 1979). Meerman (1980, p. 126) therefore concludes that "expansion of the tax system as the primary mode of financing human development may be largely foreclosed, unless the perspective is very long indeed." He suggests that complete reliance on central government revenues for financing development projects is no longer feasible and considers alternative ways of tapping the resources of the private sector, such as fees and village contributions through direct labor or services in kind, as well as measures that would reduce costs through increased efficiency and reduced wastage.

Some countries are already using alternative methods of financing, but many others would consider them an innovation. One such innovation is education vouchers, which have been advocated in the United States

(Friedman 1962) and in the United Kingdom (West 1970) as a method that allows public subsidy of education without public provision. The idea is that instead of being offered "free" education, students or their parents should be given a voucher (financed from public funds) of a certain monetary value that could be used to pay fees at schools, colleges, or other educational institutions. All institutions, both public and private, would charge fees. Students or parents could then choose between a free education at a school where fees equaled the value of the voucher, and the alternative option of supplementing the voucher for the sake of high-cost, high-fee schooling. There are many variants of this basic idea, some involving compensatory vouchers for low-income families, which would enable them to enjoy higher-than-average subsidies. Although the idea of education vouchers has been extensively discussed in the academic literature in the United States and United Kingdom, there have been few attempts to put it into practice apart from a brief experiment in Alum Rock, California (Mecklenburger and Hostrop 1972).

The idea of education vouchers has not been widely discussed in developing countries, although recent changes in the pattern and level of subsidies for universities in Chile will reduce general subsidies for higher education and introduce for the 20,000 best students special scholarships, which resemble education vouchers in some respects (Bowman, Millot, and Schiefelbein 1984). Private schools or institutions are important in some developing countries, however, and in several cases have been expanded in response to private demand and a shortage of public sector places. Pakistan, for example, reintroduced private schools because of the increasing pressure of private demand for education, and the limited growth of government expenditure on education. Private education has also expanded rapidly in the Philippines and in Korea, where households bear a significant share of the costs of secondary schooling, both in private and public institutions.

In Latin America, however, where private education is already important, some private schools are now facing a financial crisis. A recent survey of public and private financing of education noted that up to 70 percent of the private schools in Colombia might be operating at a loss and concluded, "The serious financial difficulties now faced by private schools, and the steadily decreasing proportion of students who attend them, suggest that the contribution the private sector may be expected to make in the future towards meeting the rising demand for education will be very limited" (Brodersohn 1978, p. 157).

Since public provision cannot satisfy private demand for education in many countries, it is not clear what contribution the private sector may make in the future. For political and many other reasons, it is likely to vary considerably between countries. World Bank research in this area includes a study of existing levels and patterns of private schooling in developing countries (Tan 1985b), and the Bank is now paying more attention to the role of private schools in project appraisal and education sector analysis. Furthermore, the Bank is investigating the scope for increases in tuition fees and other user charges in education. Recent analytical work on the efficiency and equity implications of cost recovery includes reviews of the theoretical issues involved in pricing policy for education and health (Jimenez 1984; Mingat and Tan forthcoming), a summary of existing policies and practice on fees and user charges (Ainsworth 1984), and empirical studies on the effects of fees on demand for education in Mali, Malawi, and Tanzania (Birdsall et al. 1983; Tan 1985a; Tan, Lee, and Mingat 1984).

All these examples reflect the growing concern, both in the World Bank and in governments of various developing countries, with the issue of cost recovery. A recent review of current policies on cost recovery in more than twenty developing countries (Ainsworth 1984) concluded that the potential for increasing cost recovery by raising fees is promising in some countries, especially for secondary schooling, and it recommended that cost recovery be improved through better enforcement of existing pricing policies. It is too early to say whether such developments will help to reduce the financial pressures already described. In some countries, fees or the growth of private education are clearly ruled out on political grounds, and alternative methods of financing may be administratively difficult. Nevertheless, many developing countries are relying more on nongovernment sources of funds and are now recognizing that cost recovery is crucial for the further expansion of educational investment in the 1980s and 1990s.

The Role of Fees

During the 1960s and 1970s most developing countries made an ideological commitment to free education. Thus tuition fees in the public sector tended to be reduced or even abolished. With the rapid increase in government expenditure that followed, however, some developing countries (for example, Malawi and Mauritius) began to consider reintroducing tuition fees, while others introduced charges for board and lodging. Recent analysis (Birdsall 1982; Thobani 1983; Mingat and Tan forthcoming) has examined the implications of increases in fees and concluded that in certain circumstances an increase in tuition fees may contribute to both efficiency and equity.

This conclusion runs contrary to the belief in most developing countries that charging tuition fees for education would be inefficient because of the existence of externalities, and inequitable because it would limit access to the rich. Because of the increasing budgetary constraint on government expenditure in developing countries, however, a system of free education will in many cases lead to excess demand for school places. As a result, some form of rationing (for example, competitive examinations) will probably be necessary, but experience has shown that it is the rich who are most likely to benefit from such a measure and the poor most likely to be excluded. Thus, the idea that free education is more efficient and equitable than tuition fees is being challenged in several countries.

Thobani (1983) suggests as a rule of thumb that whenever there is excess demand for a service, the price should be raised and additional revenues used to expand the service, up to the point where further investment is no longer socially profitable. Provided that the social rate of return can be accurately measured, this method would lead to a more efficient use of resources than rationing, since it would maximize social benefits. Thobani also suggests that it would be more equitable than rationing since the poorest pupils, or regions, are the ones likely to be denied access to schools or to suffer from low-quality schooling. This is the situation in Malawi, where an open-door admissions policy combined with low fees has led to a steady deterioration of quality, as is reflected in the high student-teacher ratios.

Fees for primary education remained unchanged in Malawi between 1975 and 1982, when they covered approximately 20 percent of total primary school recurrent expenditure. They were originally intended to cover all costs apart from teacher salaries (they were to include textbooks and other materials, for example), but beginning in 1978 the government received a credit from the International Development Association to finance textbooks. When that expired, it had the choice of increasing school fees, providing an additional K3.8 million in subsidy (which represents nearly a third of the total primary education budget), or allowing the supply of textbooks to fall.

Malawi's open-door policy for primary education meant that anyone who could pay fees between K2 or K7.5 could enroll (1K was approximately equal to US$1 in 1982). At first sight, therefore, it seems meaningless to talk of excess demand for primary school education. Thobani shows, however, that because total expenditure is limited by budgetary constraints, the result is overcrowding and low-quality schooling. The average class size at present is sixty-six; the government's aim is to reduce this to fifty. But at the present levels of subsidy the government cannot afford to satisfy the demand for primary education and also raise quality. Thobani therefore argues that for the sake of efficiency primary school fees should be raised and the additional funds used to reduce average class size and provide more books and supplies. An increase in fees could also be justified on grounds of equity since the poorest pupils are likely to suffer most from a deterioration in quality.

There is also considerable excess demand for secondary school places in Malawi. Only one in nine primary school graduates manages to gain a secondary school place, and pupils often sit the Primary School Leaving Certificate Examination more than once in order to improve their scores and increase their chances of going on to secondary school. An increase in secondary school fees might free government revenue, which could then be used to expand secondary schools or improve the quality of primary schools. Either possibility would increase efficiency and be likely to benefit the poorest pupils, who are currently denied secondary education.

This was the argument of a World Bank report on Malawi in 1981, as a result of which the government of Malawi decided in April 1982 to increase fees in both primary and secondary schools. Tuition fees were raised by 25 to 50 percent and boarding charges increased by 100 percent. This change will provide an opportunity to analyze the effects of increased levels of fees on private demand for education, and on the access of different income groups. Preliminary analysis (Tan, Lee, and Mingat 1984) suggests, however, that a moderate fee increase in Malawi is unlikely to precipitate a large dropout, particularly among secondary-school pupils currently enrolled, and that if the increased revenue is used to expand the number of school places, the fee increase could eventually lead to an increase in school enrollment. The fact is that in Malawi enrollment is limited by the supply of school places rather than by unwillingness or inability to pay fees. Excess demand clearly exists in this case, but in some developing countries it does not. Thobani's rule of thumb would not then apply.

Another study (Birdsall forthcoming) suggests that higher fees may actually cause private demand for education to increase if they are used to improve certain conditions such as distance from the nearest school and quality of schooling. (In Mali, Brazil, and Malaysia, for example, distance from school has a negative effect on school attendance, while improvements in school quality have a positive effect, particularly for children from poorer households.) Provided that additional income derived from increased fees is used to expand education in rural areas and thus increase access in remote villages, or to increase the quality of schools in these areas, an increase in fees may actually increase demand for education among poor households previously denied access or confined to low-quality schools.

Thus the likely effects of an increase in fees will probably be better understood if policymakers first analyze the factors that influence private demand for education. Although data on this are not readily available in a developing country, there may be indirect evidence of willingness to pay for education. In Mauritius, for example, both primary and secondary education have been free since 1977, yet parents continue to pay substantial sums for private tuition because of the low quality of some schools and the fierce competition for places in the best secondary schools. A recent study of financing education in Kenya (Bertrand and Griffin 1984) stresses that willingness to pay and ability to pay are far from identical in developing countries, and that more attention should be paid to the former.

The political difficulties of introducing or increasing fees at the primary and secondary level should not be underestimated, however. Some countries would also experience substantial administrative problems (and costs) in collecting fees in primary and secondary schools. Ainsworth (1984) adds that in some cases implementation costs could exceed the value of fee revenue, but this would not necessarily continue if fee levels were higher or more efficient, low-cost systems of fee collection were developed. The conclusion is that both the costs and the benefits of cost recovery for primary and secondary schooling need to be carefully weighed.

In some countries the case for increasing or introducing fees is even stronger at the university level. In many developing countries, university students pay no fees and receive free board and lodging, books, and sometimes even living allowances. In Malawi, where the cost of university education is 250 times the average cost of primary schooling, university students not only receive free tuition, meals, and accommodation, but they are given K12 a month pocket money; meanwhile, primary school pupils pay between K2 and K7.5 a year in tuition fees, and secondary school pupils pay K30 a year for tuition and between K75 and K100 a year for board (Tan, Lee, and Mingat 1984). In Burkina Faso, the private *costs* of primary education (fees plus earnings forgone) are CFA 20,000 a year, but university students receive a net *subsidy* of CFA 372,000 (Psacharopoulos 1982b).

A similar situation exists in many African countries. Mingat and Psacharopoulos (1984) show that in Francophone Africa, scholarships absorb 40 percent of the total higher education budget—a far higher proportion than that at the primary or secondary level, and ten times as high as the average in Asia. Yet the political problems involved in reducing these subsidies may be formidable. Thobani (1983) recommends increasing fees for higher education, but "not so much as to cause political riots." How much is that? In Ghana, where the government did attempt to introduce fees for board and lodging at universities in 1971, student opposition proved crucial (Williams 1974). The political problems associated with increasing fees for a powerful elite should not be underestimated, but neither should the inequity of existing systems. Another question that arises with respect to higher education is whether students should be subsidized by means of scholarships, bursaries, or loans. There is now considerable information on this issue.

The Use of Student Loans in Developing Countries

In the past twenty years many developed and developing countries have established student loan programs under which loans are provided by government agencies, commercial banks, or other financial institutions (usually with a government guarantee and some form of interest subsidy). After graduating or leaving higher education, students must repay the loan, with or without interest.

Such a system offers many advantages. Education is a profitable private investment, yet many students cannot afford to finance it out of their own or family resources. Student loans provide money when it is needed, and this can be repaid in the future when the graduate is enjoying the financial benefits of higher lifetime earnings. This system is more equitable than one in which all costs are met from public funds, since the latter involves a transfer of income from the average taxpayer to those who, in the future, are likely to enjoy higher-than-average incomes as a result of their education. Student loans have therefore been advocated for developing countries on the grounds of both equity and efficiency (Rogers 1971; Fields 1974).

Doubt has been expressed about the feasibility of using student loans in developing countries, however. Critics have argued that students will be unwilling to borrow, that loans will therefore discourage low-income students, and that the administrative problems involved in collecting repayments and preventing defaults would make loans impractical in poor countries and would wipe out any potential saving of public funds. Yet loan schemes do exist in many countries. What lessons can be drawn from international experience?

One of the first countries to introduce student loans was Colombia, where the Instituto Colombiano de Credito Educativo y Estudios Tecnicos en el Exterior (ICETEX) was established in 1950. A study of the effects of student loans in Colombia (Jallade 1974b) concluded that loans had helped to increase demand for education by reducing private costs, had enabled some poor students who could not otherwise have afforded it to enroll in higher education, but had not served to redistribute income in favor of the poor, because loan recipients often came from upper-income families. Nor had student loans been effective in shifting the financing of education from public to private sources. Because so many student borrowers were in private universities, publicly funded loans were a significant source of funds for private universities in Colombia (see table 6-10). Student loans therefore did not do much to shift the burden of financing higher education away from taxpayers to students, but were a cheap way to channel funds both to private universities and to public sector students.

The experience in Colombia shows that student loans can be used to achieve various objectives, but will not, on their own, solve the financial problems of higher education or redistribute income from the rich to the poor. The effectiveness of student loans in achieving such goals depends on policies on fees. Since there was no increase in fees in Colombia, the impact of loans on income distribution was limited.

The development of student loan schemes continued in Latin America during the 1970s, and at the same time a few small-scale schemes were initiated in some other developing countries. A short-lived experiment in Ghana provides some interesting lessons (discussed below), although the scheme was withdrawn after only a year because of a change in govern-

Table 6-10. *Student Loans and the Financing of Universities in Colombia*

Item	1969 Public	1969 Private	1970 Public	1970 Private
1. Total university revenue (thousand pesos)	707,300	210,100	776,600	269,400
2. Total revenue originating from tuition fees (thousand pesos)	39,150	146,000	37,800	187,000
3. Tuition fees as a proportion of total revenue (2/1 × 100)	5.5	69.5	4.9	69.4
4. Number of ICETEX loans	5,720	3,480	6,640	3,560
5. Average amount of loan (pesos)	4,360	6,320	4,470	6,230
6. Average amount of tuition financed per loan (pesos)	260	4,430	290	4,520
7. Total tuition fees financed through ICETEX loans (6 × 4, thousand pesos)	1,490	15,420	1,930	16,090
8. Loan-financed tuition fees as a proportion of total revenue from tuition fees (7/2 × 100)	3.8	10.6	5.1	8.6
9. Loan-financed tuition fees as a proportion of total university revenue (7/1 × 100)	0.2	7.3	0.2	6.0

Source: Jallade (1974b), table 9.

ment. Recently, however, growing budgetary constraints have led to a revival of interest in the potential of student loans as a way of shifting the financial burden of educational investment from public to private sources. One study (Woodhall 1983) that has looked at experience with student loans in developed and developing countries notes that student loan schemes now exist in eighteen countries in Latin America and the Caribbean (see table 6-11), and that there are a number of loan programs

Table 6-11. *Repayment Terms of Student Loans, Selected Countries, Latin America and the Caribbean, 1978*

| Country | Interest (percent) | | Length of repayment (years) | Grace period (months) |
	During study	During repayment		
Argentina	Linked to cost of living and Bank rate		Same as borrowing	12
Bolivia	5	5–15	Maximum 10	3
Brazil				
(APLUB)	5	10	Same as borrowing	6–12
(Caixa Economica Federal)	12	12 +	Variable	12
Chile				
(Catholic University)	Linked to cost of living		6	12
Colombia				
(ICETEX)	3–14	6–16	Variable	3–6
Costa Rica	6–8	6–8	n.a.	2–6
Dominican Republic	12	12	3 × borrowing	Variable
Ecuador	Variable		6	Variable
Honduras	8	8	8	3–6
Jamaica	6	6	9	12
Mexico				
(Bank of Mexico)	n.a.	8.5–12	7	12
Nicaragua	3	6	n.a.	12
Panama	n.a.	5	15	n.a.
Venezuela				
(Educredito)	8		Variable	6
(SACUEDO)	3–8	3–8	2 × borrowing	6–12

n.a. Not available.

Note: APLUB (Brazil) = Associação des Profissionais Liberais Universitarios do Brasil; ICETEX (Colombia) = Instituto Colombiano de Credito Educativo y Estudios Tecnicos en el Exterior; SACUEDO (Venezuela) = Sociedad Administadora de Credito Educativo para la Universidad de Oriente.

Source: Woodhall (1983).

in Africa and Asia (for example, in Kenya, Nigeria, India, Sri Lanka, and Pakistan), and one in Israel. In some countries, students receive loans for both tuition fees and living expenses, but in several cases tuition is free and loans are needed only for living expenses. The terms of repayment vary considerably, but most student loans are highly subsidized, so that graduates pay interest well below market rates.

Many of the loan schemes in Latin America and the Caribbean have been helped by financial or technical assistance from international agencies, notably the Inter-American Development Bank (IDB) and the U.S. Agency for International Development (AID). IDB loans have been given to student loan institutions in Panama, Jamaica, Trinidad and Tobago, Barbados, Honduras, and Costa Rica. Loans or grants from AID have helped establish or expand student loan programs in Brazil, Colombia, Ecuador, the Dominican Republic, Honduras, Nicaragua, and Peru.

Experience with these loan schemes shows that student loans can and do work, that students are willing to borrow, and that the existence of loans has helped to increase private demand for education and has enabled many poor students to finance their own education. Because of long repayment periods and interest subsidies, however, it takes a long time for student loans to provide a significant source of funds for higher education. Moreover, the effects of inflation can erode the value of loan repayments, unless students repay the loans in real, rather than nominal terms. As a cost-recovery mechanism, student loans are of limited value in the short run. No loan programs are yet fully self-financing; one study of student loan programs in Latin America (Herrick, Sharlach, and Seville 1974) has suggested that it would be ten or twenty years before a revolving fund could become self-financing, and even this seems rather optimistic. The Colombian ICETEX scheme has been in existence for more than twenty years, yet in 1979 loan repayments provided only 20 percent of its total income because the number of loans grew rapidly, from about 1,000 in 1963 to nearly 27,000 in 1981 (table 6-12).

Although student loans cannot promise quick savings and clearly cannot solve the financial problems threatening higher education in developing countries, there is evidence that they can contribute to a solution. In the long run, loan repayments can provide a significant source of funds. Without a system of loans, the problem may simply get worse, given the very high levels of subsidy for university students in most developing countries. Furthermore, student loan schemes can successfully involve nongovernment capital, as illustrated by the contribution of commercial banks and public or private enterprises in Colombia (see table 6-12). In Brazil, donations from private organizations and trade unions are important, and in Sri Lanka and Pakistan, commercial banks finance and administer student loan schemes.

Table 6-12. *Sources of Finance for ICETEX Scheme,*
Colombia, 1979

Source	Income (thousands of dollars)	Percent
Government budget	236,340	32.6
Administration of		
enterprise funds	132,752	18.3
Bank loans	140,000	19.3
Central Bank	35,156	4.9
IDB	34,000	4.7
Loan repayments	146,832	20.2
Total	725,080	100.0

Source: Woodhall (1983), p. 39.

The capacity of student loans to shift the financing burden from public to private sources depends, however, on the level of fees. If student loans are introduced in conjunction with an increase in fees, the impact on public funds is obviously greater than if tuition or boarding costs remain highly subsidized. The government of Ghana attempted to introduce fees for board and lodging together with loans for university students in 1971. The scheme was unpopular with students, since it meant their private costs would go up. After a change of government in 1972 the scheme was withdrawn. This experience is sometimes quoted as proof that loans are not feasible in Africa because of political difficulties. Yet an evaluation of the experience in Ghana suggests that the failure was partly owing to inadequate preparation and the absence of a sustained campaign to explain the policy to the public. It has also been argued that the loan scheme was "an accidental victim of the political circumstance of the change of government, with abolition of the scheme being a useful tactical weapon for the new government in the early days" (Williams 1974, p. 342).

Although there are likely to be formidable political difficulties in reducing subsidies for higher education either by student loans or increased fees, it is not necessarily impossible. Experience in both developed and developing countries shows that student loans are not a panacea, but that they can contribute to both efficiency and equity goals, are flexible as a means of finance, and can help to increase the share of private finance for educational investment. Their impact will be greater, however, if they are combined with other financial reforms, including measures to reduce costs and increase efficiency (see chapters 7 and 8).

The Financing of Vocational
and Technical Education and Training

Vocational education and training for agricultural, industrial, or commercial occupations are provided and financed in a variety of ways, but recently employers have become increasingly involved in the provision or the financing of training opportunities for workers. The training may take place in vocational schools, colleges, or other institutions, or may be acquired on the job in the case of apprentices or experienced workers. Combinations of on-the-job and off-the-job training can also be found. Whatever the administrative structure in a country, the role of employers in providing or contributing to training is increasingly recognized and emphasized.

Because of a number of concerns, however, new methods of organizing and financing vocational education and training are being developed. Governments are seeking to transfer some of the financial burden of education from public to private funds, and are also attempting to ensure that vocational education is relevant to the needs of the labor market. It is thought that if employers are directly involved in planning the curriculum or in providing the training, then vocational education will be more relevant to labor market demands and will develop industrial skills more effectively.

The choice between on-the-job and off-the-job training, apprenticeship, vocational schools or skill center training, day or block release, or other methods of combining practical experience and technical instruction must take into account many factors, including relative costs, the cost-effectiveness of different methods of training, the availability of skilled instructors and equipment, and production processes and other characteristics of the labor market. Basically, the costs of training have to be shared by three parties: the government, which provides funds for vocational education or training from general or earmarked taxation or other sources of government revenue; employers, who may provide training directly, may finance it through payment of general or specific taxes (such as payroll or turnover tax or training levies), and may also pay wages or salaries to trainees; and the trainees, who may pay fees for vocational education or training, or may work for reduced wages while being trained and thus bear the opportunity costs of training through earnings forgone.

Investment in training generates both costs and benefits to the individual worker or trainee, to the employer, and to society as a whole. The rate of return to training has been analyzed in the same way as the rate of return to investment in general education (Mincer 1962; Becker 1975), but there has been less empirical work on training programs in developing countries. Two reviews (Zymelman 1976; Metcalf 1985) have indicated that the social, corporate, and private returns to vocational training in developing countries seem high enough to justify expanding training, but that no general conclusions can be drawn on the relative merits of on- and off-the-job training. Because the costs per student in vocational training institutions are often high, their training programs will be profitable only if the courses are shorter and more efficient than formal education or on-the-job training. A study of the National Training Service (SENATI) programs in Peru (Psacharopoulos 1982a), for example, showed that short courses had a much higher rate of return than long-term apprenticeship programs.

Vocational training in some developing countries is financed jointly by government and by employers through payroll taxes or training levies. There are many examples in Latin America, including the National Service for Apprenticeship (SENA) in Colombia, the National Service for Industrial Apprenticeship (SENAI) in Brazil, and SENATI in Peru (see table 6-13). Almost all vocational training insitutions in Latin America are financed by some form of payroll tax, which means that employers are required to make compulsory contributions to the costs of industrial, technical, and vocational training in specialized institutions. This is one important way in which employers, both private and public, can share directly in the financing of training. Some other developing countries are seeking to develop similar institutions.

According to Ducci (1983), there were twenty-three vocational training institutions in Latin America in 1980, and the total costs of all their training activities was about US$1 billion. A study of the financing of vocational training in Latin America (Kugler and Reyes 1978) concluded that the payroll tax has proved to be an effective financing mechanism since it has permitted the creation and growth of training institutions that are a workable alternative to more traditional systems of technical instruction within the schools and on-the-job training, which may be inefficient in view of the average size of industrial establishments in Latin America. However, care must be taken in devising a suitable form of payroll tax so as not to discourage small-scale employers; in some circumstances, a payroll tax might increase unemployment.

An important source of funds for training in some cases is bilateral and multilateral aid, which may finance specific skill training as part of an investment project in transport, irrigation, or some other sectors. The general conclusion from recent experience with training in developing countries is that practical on-the-job training, either as a component of project investment or as an investment by employers in the skills of their workers, is an important and profitable form of investment that can complement formal education, help to overcome financial constraints, and increase the profitability of other forms of investment.

Community Involvement in Financing Education

As noted earlier, some countries have attempted to overcome financial constraints by using direct labor to build schools, by allowing communities to provide goods and services in kind rather than cash payments, and by relying on other forms of local community involvement or self-help. Zymelman (1973) concludes that self-help has great potential as an educational financing method because it can provide extra resources and ensure that they are used effectively and flexibly.

In some countries local communities are responsible for building schools and use labor and local materials donated by the community. In Nepal, for example, almost all primary and many secondary schools are built and maintained by local communities. This practice significantly eases pressure on the government's capital and maintenance budget. In Kenya, the harambee schools are another example of self-help. In Tanzania, villagers construct primary school buildings and teachers' houses but the government provides the construction materials. In Malawi, self-help construction of primary school buildings is estimated to cost only a third of the amount it takes to construct conventional buildings. In some cases, teachers and pupils contribute to the costs of education by producing and selling goods in the school. Citing evidence from Trinidad and Tobago, the Dominican Republic, Panama, Honduras, and Cuba, Brodersohn (1978) concludes that between 25 and 50 percent of the operating and maintenance costs of a school can be financed by the sale of goods produced in the school; between 15 and 30 percent of recurrent costs can be financed with community inputs; and between 15 and 30 percent of capital expenditure can be financed with community inputs. An ambitious program of rural education is unlikely to be able to rely solely on local financial contributions, however. The experience of the MOBRAL mass literacy program in Brazil, which benefited 17 million people between 1970 and 1974, suggests that an extramural mass literacy program requires substantial government funds; in this case, 95 percent of the financing came from sports lottery receipts and income tax revenue (IDB 1978, p. 155).

The local community may contribute by providing financial support as well as goods or services in kind. A review of alternative financing methods (Meerman 1980) suggests that a rural primary school system

Table 6-13. Sources of Financing of Vocational Training Institutions, Latin America, 1975

| Country | Insti-tution | Year of estimate | Public budget | External assist-ance | Payroll tax rate (percent) | | | | | Other sources |
					Govern-ment	In-dustry	Ser-vices	Agricul-ture	Tax base[a]	
Argentina	CONET	1959	Yes	—	—	10	—	—	All	—
Bolivia	FOMO	1972	Yes	Yes	—	—	—	—	—	Yes[b]
Brazil	SENAI	1942	—	—	—	10	—	—	All[c]	—
Brazil	SENAC	1946	—	—	—	—	10	—	All	Yes[d]
Chile	INACAP	1966	Yes[e]	—	—	—	—	—	—	Yes[f]
Colombia	SENA	1957	—	—	5	20	20	20	10 + [g]	—
Costa Rica	INA	1965	—	—	—	10	10	10	5 + [g]	—
Ecuador	SECAP	1966	Yes	—	—	5	5	—	All	Yes[h]
Guatemala	INTECAP	1972	Yes	—	—	5	5	5	5 + [i]	—
Honduras	INFOP	1972	—	—	5	10	10	10	5 + [g]	—
Mexico	ARMO	1965	Yes	Yes[j]	—	—	—	—	—	—
Nicaragua	INA	1967	Yes	Yes	—	—	—	—	—	—

Panama	Labor Ministry	1965	Yes	—	—	—	—	—	—	—
Paraguay	SNPP	1971	Yes	—	—	10	10	10	All	—
Peru	SENATI	1961	—	—	—	15	—	—	15+	—
Uruguay	UTU	1942	Yes	—	—	—	—	—	—	—
Venezuela	INCE	1959	Yes^k	—	—	20	20	20	5+	Yes^l

— Blank cells mean that a particular source was not used.

Note: Covers only sources accounting for more than 2½ percent of the total budget of the institution.

a. Refers to the minimum number of workers that an enterprise must have in its employ to be subject to the payroll tax.

b. Income of business entities; was 6 percent of the total in 1975.

c. There is a 0.2 percent surtax on industrial firms employing more than 500 workers.

d. Income of trainee workers; was 10.8 percent of the total in 1975.

e. Through the budget of the Development Corporation (CORFO).

f. Contributions of trainee workers; was 19.7 percent of the total in 1975.

g. Autonomous institutes and state enterprises are taxed as private firms.

h. A 2 percent levy on machinery imports.

i. In the agricultural sector the tax is paid by firms with 10 or more workers.

j. 38 percent of the budget in 1965–72 (UN).

k. 20 percent of the payroll tax.

l. 0.5 percent of the wages paid to trainee workers.

Source: Kugler and Reyes (1978).

161

could be developed in which the central government provided the administrative structure—that is, hired the teachers, procured the books, and designed the curriculum—while the local community paid the costs of teachers, building, equipment, and books. Local communities already cover the costs of one or more of these items in many countries, and the degree and variety of village participation in some countries demonstrates that educational investment need not rely entirely on central government funding.

Experience shows that there is a wide variety of ways of financing educational investment, and that governments can, if they are sufficiently determined, devise mechanisms for shifting part of the financial burden of education to individual students and their families (for example, through tuition fees or student loans), to employers (through training levies and payroll taxes), or to local communities (through self-help building or help with operating costs). In addition, they may be able to devise new taxes earmarked for education such as a graduate or professionals' surtax of the type proposed in several Latin American countries. Much more attention is now being paid to such questions in developing countries as governments recognize that central government funding is not the only, or necessarily the most desirable, way to finance all educational investment.

References

Ainsworth, M. 1984. User Charges for Social Sector Finance: Policy and Practice in Developing Countries. Washington, D.C.: World Bank, Country Policy Department.

Armitage, J., and R. H. Sabot. Forthcoming. Efficiency and Equity Implications of Subsidies of Secondary Education in Kenya. In *Modern Tax Theory for Developing Countries*, ed. David Newbery and Nicholas Stern. New York: Oxford University Press.

Becker, Gary. 1975. *Human Capital: A Theoretical and Empirical Analysis, with Special Reference to Education.* 2d ed. New York: National Bureau of Economic Research.

Bertrand, T. J., and R. Griffin. 1984. The Economics of Financing Education: A Case Study of Kenya. Washington, D.C.: World Bank, Country Policy Department.

Birdsall, Nancy. 1982. Strategies for Analyzing Effects of User Charges in the Social Sectors. World Bank Country Policy Department. Discussion Paper no. 1983-9. Washington, D.C.

———. Forthcoming. The Impact of School Availability and Quality on Children's School Attainment in Brazil. *Journal of Development Economics.*

————, et al. 1983. Demand for Primary Schooling in Rural Mali: Should User Fees Be Increased? World Bank Country Policy Department Discussion Paper no. 1983-8. Washington, D.C.

Blaug, Mark. 1982. The Distributional Effects of Higher Education Subsidies. *Economics of Education Review* 2, no. 3 (Summer):201–31.

Bowman, Mary-Jean, Benoit Millot, and Ernesto Schiefelbein. 1984. The Political Economy of Public Support of Higher Education: Studies in Chile, France, and Malaysia. Washington, D.C.: World Bank, Education Department.

Brodersohn, Mario. 1978. Public and Private Financing of Education in Latin America: A Review of Its Principal Sources. In *The Financing of Education in Latin America*. Washington, D.C.: Inter-American Development Bank.

Ducci, Maria Angelica. 1983. *Vocational Training: An Open Way*. Montevideo: CINTERFOR (International Labour Office).

Eicher, Jean Claude. 1984. *Educational Costing and Financing in Developing Countries: Focus on Sub-Saharan Africa*. World Bank. Bank Staff Working Paper no. 655. Washington, D.C.

Fields, Gary S. 1974. Private Returns and Social Equity in the Financing of Higher Education. In *Education, Society and Development: New Perspectives from Kenya*, ed. D. Court and D. P. Ghai. Nairobi: Oxford University Press.

————. 1975. Higher Education and Income Distribution in a Less Developed Country. *Oxford Economic Papers* 27, no. 2 (July):245–59.

Friedman, M. 1962. *Capitalism and Freedom*. Chicago: University of Chicago Press.

Hansen, W. Lee, and Burton A. Weisbrod. 1969. The Distribution of the Costs and Benefits of Public Higher Education: The Case of California. *Journal of Human Resources* 4 (2):176–91.

Herrick, Allison, Howard Sharlach, and Linda Seville. 1974. *Intercountry Evaluation of Education Credit Institutions in Latin America*. Washington, D.C.: U.S. Agency for International Development.

Inter-American Development Bank (IDB). 1978. *The Financing of Education in Latin America*. Washington, D.C.

Jallade, Jean-Pierre. 1973. *The Financing of Education: An Examination of Basic Issues*. World Bank Staff Working Paper no. 157. Washington, D.C.

————. 1974a. *Public Expenditures on Education and Income Distribution in Colombia*. Baltimore, Md.: Johns Hopkins University Press.

————. 1974b. *Student Loans in Developing Countries—An Evaluation of the Colombian Performance*. World Bank Staff Working Paper no. 182. Washington, D.C.

Jimenez, E. 1984. Pricing Policy in the Social Sectors: Cost Recovery for Education and Health in Developing Countries. Washington, D.C.: World Bank, Country Policy Department.

Kugler, B., and A. Reyes. 1978. Financing of Technical and Vocational Training in Latin America. In Inter-American Development Bank. *The Financing of Education in Latin America*. Washington, D.C.

Mecklenburger, J. A., and R. W. Hostrop, eds. 1972. *Education Vouchers: From Theory to Alum Rock*. Homewood, Ill.: ETC Publications.

Meerman, Jacob. 1979. *Public Expenditure in Malaysia: Who Benefits and Why?* New York: Oxford University Press.

———. 1980. Paying for Human Development. In *Implementing Programs of Human Development*, ed. Peter T. Knight. World Bank Staff Working Paper no. 403. Washington, D.C.

Meesook, Oey Astra. 1984. *Financing and Equity in the Social Sciences in Indonesia*. World Bank Staff Working Paper no. 703. Washington, D.C.

Metcalf, David H. 1985. *The Economics of Vocational Training: Past Evidence and Future Considerations*. World Bank Staff Working Paper no. 713. Washington, D.C.

Mincer, Jacob. 1962. On-the-Job Training: Costs, Returns and Some Implications. *Journal of Political Economy* 70, no. 5, part 2 (October):50–79.

Mingat, A., and G. Psacharopoulos. 1984. Education Costs and Financing in Africa: Some Facts and Possible Lines of Action. Washington, D.C.: World Bank, Education Department.

Mingat, A., and J. P. Tan. Forthcoming. Expanding Education through User Charges in LDC's Austerity: What Can Be Achieved? *Economics of Education Review*.

Pechman, J. A. 1970. The Distributional Effects of Public Higher Education in California. *Journal of Human Resources* 5, no. 3 (Summer):361–70.

Psacharopoulos, George. 1977. The Perverse Effects of Public Subsidisation of Education. *Comparative Education Review* 21, no. 1 (February):69–90.

———. 1982a. Peru: Assessing Priorities for Investment in Education and Training. Washington, D.C.: World Bank, Education Department.

———. 1982b. Upper Volta: Is It Worth Spending on Education in a "High-Cost" Country? Washington, D.C.: World Bank, Education Department.

Rogers, Daniel. 1970. The Economic Effects of Various Methods of Educational Finance. World Bank Economics Department Working Paper. Washington, D.C.

———. 1971. Financing Higher Education in Less Developed Countries. *Comparative Education Review* 15 (February):20–27.

Selowsky, Marcelo. 1979. *Who Benefits from Government Expenditure? A Case Study of Colombia*. New York: Oxford University Press.

Tait, Alan. 1979. International Comparisons of Taxation for Selected Developing Countries 1972–76. *IMF Staff Papers* 26, no. 1 (March).

Tan, J. P. 1985a. The Private Direct Cost of Secondary Schooling in Tanzania. *International Journal of Educational Development* 5, no. 1:1–10.

———. 1985b. Private Enrollment and Expenditure on Education: Some Macro Trends. *International Review of Education* (May).

Tan, Jee-Peng, Kiong Hock Lee, and Alain Mingat. 1984. *User Charges for Education: The Ability and Willingness to Pay in Malawi*. World Bank Staff Working Paper no. 661. Washington, D.C.

Thobani, Mateen. 1983. *Charging User Fees for Social Services: The Case of Education in Malawi*. World Bank Staff Working Paper no. 572. Washington, D.C.

West, E. G. 1970. *Education and the State*. 2d ed. London: Institute of Economic Affairs.

Williams, Peter. 1974. Lending for Learning: An Experiment in Ghana. *Minerva*. 12 (3):326–45.

Woodhall, Maureen. 1983. *Student Loans as a Means of Financing Higher Education: Lessons from International Experience*. World Bank Staff Working Paper no. 599. Washington, D.C.

World Bank. 1980. *Education*. Sector Policy Paper. Washington, D.C.

———. 1982. *IDA in Retrospect: The First Two Decades of the International Development Association*. New York: Oxford University Press.

Zymelman, Manuel. 1973. *Financing and Efficiency in Education: Reference for Administration and Policy Making*. Cambridge, Mass.: Harvard University Press.

———. 1976. *The Economic Evaluation of Vocational Training Programs*. Baltimore, Md.: Johns Hopkins University Press.

———. 1982. *Educational Expenditures in the 1970s*. Washington, D.C.: World Bank, Education Department.

7

The Costs of Education

Because of the increasing financial constraints on educational investment, developing countries are not only searching for alternative ways of financing education, but they are also paying closer attention to the costs of educational investment and attempting to reduce unit costs by improving efficiency. This chapter looks at different ways of defining, measuring, and analyzing educational costs and at the methods used for estimating future costs of educational investment and for identifying their determinants in order to reduce and control them. The question of efficiency is discussed in chapter 8.

Alternative Ways of Measuring and Analyzing Educational Costs

Decisions about educational investment depend on judgments about the balance between costs and benefits (see chapter 3). These judgments in turn are based on a systematic comparison of the economic benefits of education and its opportunity cost, which is measured not by actual monetary expenditure, but by the alternative opportunities forgone when scarce resources are invested in education. In developing countries, where educational investment is financed by and large from government revenue, the alternative opportunities forgone when a new school is built may be an irrigation project, a fertilizer plant, or agricultural investment or transport.

Because educational investment involves both social and private opportunity costs, government choices must take into account public or fiscal costs as well as the wider social costs. Thus it is important to specify which concept of cost is relevant to the type of decision that is to be made. Furthermore it is important to establish the appropriate method of measuring the cost. For many purposes unit costs are needed, but whereas average cost per pupil or student is relevant for cost comparisons or projections, the marginal or incremental cost of additional students may be more important when choosing whether to expand existing facilities or build new schools. The cost per pupil or student may, indeed, not be the most appropriate way of measuring unit costs. Where dropout or repetition rates are high, the cost per graduate or school completer may be more relevant than average cost per student. Decisions about alternative educational technologies require information on unit costs (such as costs per hour) and a full understanding of the cost implications of alternative technologies (such as radio or television), which requires detailed analysis of fixed and variable costs.

The considerable literature that has grown up around cost analysis offers alternative definitions and concepts of cost, each relevant to different types of decision. In addition, different specialists have defined costs in different ways; in particular, many accountants and economists differ in the way they classify costs. As a result, considerable confusion has arisen over the methodology of cost analysis, and in many cases costs have been underestimated or the cost comparisons are misleading.

The World Bank, Unesco, and IIEP have together developed methodological guidelines for the estimation and analysis of costs in education projects and have commissioned several case studies of educational costs (which are reviewed in this chapter). A number of useful methodological handbooks on cost analysis (for example, Fisher 1971; Jamison 1977; Jamison, Klees, and Wells 1978; Levin 1983) have also appeared in recent years. All this work reflects the increased interest in cost analysis and the general belief that the methods of defining and measuring costs need to be clarified.

Different types of decision require different measures of cost and different analytical techniques. The techniques of cost-benefit analysis, as we have already seen, are appropriate for evaluating the economic profitability of alternative investment projects. Cost-effectiveness analysis, however, is more appropriate for assessing the noneconomic effects of education. Whereas cost-benefit analysis compares the social or private opportunity costs of an investment with the expected monetary benefits or returns, cost-effectiveness analysis is concerned with the outcomes of education that are measurable in nonfinancial terms. Cost-effectiveness analysis is therefore appropriate for comparing alternative ways of achieving the same result, for example the development of reading or language skills. The most cost-effective technique is the one that produces the desired result at minimum cost or produces the largest gains in achievement for a given cost. Whether educational outcomes are measured in monetary or nonmonetary terms, the costs must be measured accurately.

Other types of cost analysis can be used in making educational decisions. If it is necessary to compare costs over time, for example, costs measured in current prices will have to be distinguished from costs measured in real terms (that is, in constant prices, or in terms of money of constant purchasing power). If the goal is to understand the determinants of costs, other types of cost analysis are useful; for example, cost functions can help to throw light on the way total or average costs change in relation to the size of an institution or system. The question of whether economies of scale are significant in education is important for decisions about the cost implications of expansion or contraction. Cost functions are also useful for analyzing the costs of new educational media. This chapter provides examples of all these types of cost analysis.

Some critics who have called for more detailed and systematic analysis of educational costs have pointed to a lack of cost consciousness within the World Bank itself. One report by an external advisory panel on education to the World Bank (1978, p. 11) noted, for example,

> In our field visits and reading of Bank documents, we found few attempts to make cost-effectiveness assessments of alternative projects. . . . Attention to cost analysis in connection with Bank loans for education and training appears to have been a relatively recent development. Cost analysis is still too rarely used in decisions on which projects to support. Costs are of course considered as constraints within a project, but too seldom in cost-effectiveness comparisons of alternative projects. The exception, an important one, is the growth in the Bank as elsewhere of cost-effectiveness analysis of the internal efficiency of experimental programs in education.

Because of the interest in cost analysis, educational cost studies have increased in number, particularly within the Bank, Unesco, and IIEP. Some of these studies are concerned with the costs of alternative educational techniques such as radio and television—sometimes collectively described as distance-teaching. A recent review (Perraton 1982) explores the potential of distance-teaching methods in reducing educational costs, while some other recent studies provide detailed discussions of the costs of new educational media (Jamison 1977; Unesco 1977, 1980; Eicher et al. 1982). Another interesting report contains case studies of educational cost analysis (IIEP 1972), which have also been summarized by Coombs and Hallak (1972). Furthermore, work has begun on the determinants of education costs in a sample of developing countries (Tibi 1983).

Despite the greater emphasis on cost issues, Eicher not long ago was still prompted to say, "We know much less about costs of education than we often think we do" (Eicher 1984, p. 1). The problem is that budgetary data are often inadequate for a detailed study of costs since they cover expenditures rather than real resource, or opportunity, costs. Moreover, they often present planned or provisional budget estimates rather than actual expenditure. Another problem is that the classification of recurrent and capital expenditure is often confused, and indeed the treatment of capital expenditure varies considerably from one study to another. There is very little information on private expenditure, while the data on public expenditure are often inaccurate and classifications and definitions vary between countries. International comparisons can therefore be misleading.

Some of the confusion stems from the methodologies used, as can be seen in international comparisons of average costs, which often do not make clear whether weighted or unweighted, arithmetic or geometric averages are used. As an example, Eicher compares two estimates of average costs in higher education by using Unesco data for total expenditure and enrollments in both cases; in one case, however, the geometric, rather than arithmetic average was calculated.[1] The difference between the two sets of estimates is striking and may lead to opposite policy conclusions. In fact, very rough or unreliable cost estimates are sometimes used as a basis for conclusions and policy recommendations on education investment. Individual governments and international organizations, including Unesco and the World Bank, are therefore becoming increasingly cautious about accepting past cost estimates and are devoting time and resources to gathering reliable cost information and to comparing different sources. At the same time, more attention is being given to the determinants of costs (such as economies of scale, or the influence of age structures on teacher salaries), so that the behavior of costs can be better understood and thus unit costs reduced and cost-effectiveness increased. This chapter summarizes the progress that has already been made toward a better understanding of costs.

Alternative Cost Concepts

The way in which costs are classified varies considerably. The World Bank and Unesco have attempted to standardize definitions and ways of measuring the costs of new media (Unesco 1977, 1980; Eicher et al. 1982), but have not yet standardized methods of classifying these costs. One possibility is to classify them on the basis of technical function; for example, separate costs are incurred for design and production, reproduction, distribution, and implementation of media systems. This method is appropriate for a technical analysis, but not for an economic analysis, which may be concerned with variations with respect to size (that is, fixed and variable costs) or with how costs are financed.

Thus, the classification of educational costs raises the following economic, financial, and institutional as well as technical questions:

- What has to be sacrificed? (economic issues)
- When does money have to be paid? (financial issues)
- Who has to pay? (institutional issues)
- What is the function of the inputs? (technical issues)

Because these issues are complex and differ between countries, there is no internationally standard terminology for or classification of educational costs. General agreement has been reached, however, concerning certain basic distinctions.

The distinction between *opportunity cost* and *expenditure* has already been mentioned. Any attempt to measure the economic profitability of investment must consider not only money expenditure, but also real resources that have alternative uses. To estimate the opportunity cost of resources in terms of the alternative opportunities that are sacrificed when resources are invested in one project rather than another, it is necessary to include not only the monetary value of resources (teachers' salaries, for example), but also the estimated value of resources that are not bought or sold. Students' time, for example, is usually measured by earnings forgone, and some measure may be made of the annualized value of land donated by the local community. These items do not appear in any education budget, but they represent real resources that have alternate uses.

If opportunity costs are to be estimated with any degree of accuracy, *shadow prices* may have to be used to measure the true economic value of resources when their market prices are distorted, for example, through government control of exchange rates or wages and salaries. Since data on actual expenditure on education in developing countries are often inadequate, estimates of opportunity costs and shadow prices are necessarily approximate. Even so, an effort must be made not only to analyze budgetary costs, but also to define costs as broadly as possible and to estimate the true value of all resources committed to a project.

Budget data normally distinguish between *recurrent* and *capital* expenditure. Conceptually, the distinction is straightforward. Recurrent expenditure, as the term implies, recurs regularly and covers expenditure on goods and services that bring immediate and short-lived benefits. Thus, expenditure on consumable goods, such as materials and teachers' salaries, is classified as recurrent expenditure. Capital costs or expenditure includes the purchase of durable assets, such as buildings or equipment, that are expected to yield benefits over a longer period. There are, however, many intermediate items such as textbooks that may be difficult to describe as short-lived or long-lived, consumable or durable. The usual convention is to make one year the accounting period, and thus goods or services used up within one year are regarded as recurrent costs. In practice, the crucial distinction is often not the expected life of the item, but the way it is financed. Recurrent expenditure is usually financed out of current income, or revenue, while capital expenditure is financed by loans, including loans from international agencies such as the World Bank.

The distinction between recurrent expenditure that brings short-term benefits and capital expenditure that produces long-term benefits leads to certain problems when it is applied to educational investment. All educational expenditure—both recurrent and capital—can be regarded as a means of forming human capital that will yield benefits throughout the working life of an educated person, which will last perhaps forty or fifty years. Teachers' salaries, which account for more than half of the total costs of education, are classified as recurrent expenditure, but teachers help create skills that will last for a lifetime. In other words, there is an important conceptual difference between recurrent expenditure in the accounting sense of the term, and expenditure that creates a capital asset in the economic sense of the term.

At times of budgetary stress, governments are often concerned to achieve a quick payoff and invest in projects with low recurrent costs. If most educational expenditure is regarded as a recurrent cost, education will appear unprofitable. This line of reasoning has two important implications. One is that if governments and international financing agencies recognize that educational expenditure produces long-run, rather than short-run benefits, then they may be willing to finance a greater share of expenditure through loans. Second, if this time horizon is recognized, governments may choose to give greater priority to the formation of long-term capital rather than to short-term reductions in recurrent expenditure. In practice, the distinction between capital and recurrent expenditure on education is in any case difficult to apply. It is common, for example, to treat the initial purchase of books and furniture as a capital expense, and subsequent replacements (together with repairs and maintenance) as a recurrent cost. Rather than treating the purchase of a major item of equipment as a capital cost in a single year, however, it may be more helpful to calculate the *annualized value* of capital expenditure by calculating *amortization* and *depreciation*.

A number of problems can arise in the treatment of capital costs because different projects have adopted different conventions in dealing with capital and recurrent costs, have made different assumptions about the expected life of equipment, and have used different methods of calculating amortization. Such differences can be crucial when projects involving new media have to be evaluated. In such cases, the initial capital costs are high, but future recurrent costs (in the form of repair, maintenance, and replacement of equipment) are also substantial. Serious problems have arisen in the evaluation of educational investment projects because the recurrent cost consequences of a capital investment were underestimated.

One reason the costs of new media are often underestimated is that unrealistic assumptions are made about economies of scale. To investigate whether there are economies of scale, it is necessary to distinguish between *average* and *marginal* costs, and also between *fixed* and *variable* costs. Average cost, often called unit cost, simply represents total expenditure or cost divided by the total number of students or pupils.[2]

For many purposes, a simple unweighted average may be sufficient, but in some cases it can be misleading. It would be misleading, for example, to calculate the average expenditure per student simply by adding together expenditure on primary, secondary, and higher education and then dividing this by the total number of students, regardless of the relative proportions in each level. In such a case, a *weighted average* that applied the relative proportions of enrollments as weights would be more appropriate, particularly if the proportions are likely to change over time.

Similarly, where there are substantial differences in the costs of different subjects, a weighted average would use relative costs as weights. In many countries, the funding of institutions is based on some kind of unit cost formula. If there are substantial differences between types of school, level, or subject, the formula should be based on weighted averages, and weights should accurately reflect actual cost differences. In such cases, the choice of weights becomes crucial.

Another question that will have to be considered is whether *average* or *marginal* costs are more appropriate. Strictly speaking, the marginal cost of one unit of output is the extra expenditure incurred when one additional unit is produced and the result is a marginal increase in total output. Thus, the marginal cost of enrolling an additional pupil is the extra expenditure incurred when total enrollment is increased by one.[3] In fact, it is more usual to estimate the increase in total expenditure incurred as a result of an *increment* (which may be a small group) or an additional class, rather than a single student. Thus some writers prefer the term *incremental cost*, rather than marginal cost.

In practice, it is often extremely difficult to identify marginal costs, but the concept is nevertheless important since marginal costs indicate the cost consequences of expanding the system, whereas average costs indicate the amount of money or resources devoted to each student in the existing system. Whether marginal costs are equal to average costs depends on the degree of *utilization* of resources in the existing system. If there is spare capacity, then it would be possible to increase enrollments without incurring substantial expenditure; there would be *economies of scale*, and marginal costs would be below average costs. If existing facilities are overcrowded, however, there may be *diseconomies of scale*, and marginal costs may exceed average costs.

The concept of a *cost function* can be used to investigate the relationship between average and marginal costs. The form of the cost function is partly determined by the relationship between fixed and variable costs. This distinction depends upon whether costs vary with the level of output or, in the case of education, the number of students enrolled.

The distinction between fixed and variable costs is not the same as between capital and recurrent costs, since some recurrent costs (for example, the salaries of central administrative staff) may not vary with respect to student numbers, whereas others will. In the short run, many costs may be regarded as fixed, because they do not vary with respect to small changes in the level of output. In the long run, however, or for large-scale changes in the level of output, costs that were fixed in the short run will become variable. In the very long run, all costs are variable.

Fixed and variable costs must be identified in capital-intensive projects, such as radio or television. The cost of installing a broadcasting network, for example, is fixed and independent of the number of students using the system, whereas the cost of face-to-face instruction or correspondence teaching material will vary directly with the number of students.

This brief review of alternative cost concepts shows that there is no simple answer to the question, What is the cost of education? It depends on the type of decision to be made. The first step that should be taken in answering this question is to establish the cost to whom. Social costs represent the cost to the economy as a whole, which is relevant for an assessment of education as a social investment; private costs represent the cost to the individual, which helps to determine private demand; whereas the cost to public funds is relevant to an assessment of the fiscal consequences of educational investment.

The second step is to determine whether average or marginal costs are relevant. This will depend on whether the investment decision involves a choice between different types of institution, for example, in which case average cost per student would be relevant, or a choice between expanding or keeping existing systems, in which case marginal costs would be relevant. The relationship between average and marginal costs is discussed later in this chapter. We deal first with trends in total costs, the importance of teacher salaries, and the relationship between current and capital costs.

Table 7-1. *Factors Influencing Expenditures on Primary Education, East Africa*
(percent)

Country	Primary-level expenditures ÷ GNP	Primary-level unit cost ÷ GNP per capita	Enrollment ratio	Demographic burden factor	Percentage deviation from average		
					Unit cost ÷ GNP per capita	Enrollment ratio	Demographic burden factor
Botswana	3.65	20.85	96.05	18.25	10.2	32.3	12.2
Burundi	1.26	34.09	23.44	15.81	80.2	-67.7	-2.8
Kenya	3.29	14.70	124.13	18.02	-22.3	70.9	10.8
Lesotho	1.69	10.20	108.72	15.22	-46.1	49.7	-6.4
Madagascar	2.16	12.73	107.16	15.84	-32.7	47.6	-2.6
Mauritius	2.41	16.72	100.02	14.44	-11.6	37.7	-11.2
Rwanda	3.08	33.92	54.55	16.66	79.3	-24.9	2.4
Somalia	3.13	44.41	50.05	14.08	134.7	-31.1	-13.4
Sudan	2.50	34.58	45.15	16.04	82.8	-37.8	-1.4
Swaziland	1.26	7.38	104.37	16.31	-61.0	43.7	0.3
Tanzania	3.32	19.61	100.41	16.88	3.6	38.3	3.8
Uganda	0.89	11.69	46.46	16.45	-38.2	-36.0	1.2
Zambia	2.23	13.57	97.22	16.92	-28.3	33.9	4.0
Zimbabwe	2.87	27.76	59.66	17.35	46.7	-17.8	6.7
Average for East Africa	2.23	18.92	72.62	16.26	—	—	—

Source: Zymelman (1982a), p. 13.

174

Trends in Total Costs

During the 1960s and 1970s total educational expenditure increased at a remarkable rate, both in money terms and in real terms, that is, in terms of constant purchasing power. Total public expenditure on education rose as a proportion of national income and of total public expenditure in both developed and developing countries during much of this period, although there has been a slowing down in many countries since 1975. These two ratios can be regarded as indexes of educational effort and fiscal effort. Countries can be compared on the basis of these indexes of effort as well as the following three variables: unit costs as a proportion of GNP per capita; enrollment ratios; and school-age population, as a proportion of total population, or the demographic burden (Zymelman 1976, 1982a).

The amount a country or region devotes to education does not depend on the level of economic development, as measured by GNP per capita, but is influenced by unit costs as a ratio of GNP per capita and the enrollment ratio. The relative importance of these factors varies between countries, however, and also between regions. In East Africa, for example, the proportion of GNP devoted to primary education in the late 1970s was higher in Kenya than in East Africa as a whole and unit costs as a proportion of GNP were below average, but the enrollment ratio was well above average and this clearly influenced total expenditure (see table 7-1). In the case of Burundi, however, the enrollment ratio was well below average, but unit costs were well above average.

Further analysis of educational expenditure in Africa (Eicher 1984; Lee 1984) provides additional information on the causes of high levels of educational spending in the region. Eicher (1984) concludes from an analysis of data for 122 countries that between 1960 and 1976 the enrollment ratio for the 6-to-11 age group is the most important variable to explain variations in educational expenditure as a proportion of GNP. Nevertheless, the demographic pressure is such that even if the proportion of GNP devoted to educational expenditure continues to rise until the end of the century, universal primary education (UPE) will still not be achieved in either Africa or Asia. Thus, Eicher (1984, p. 56) concludes that "governments *do* have good reasons to be concerned about the rising trend of total costs and about their ability to finance those costs in the future. . . . The need for cost-reducing measures and more generally for policies towards cost-effectiveness is everywhere present and is getting more urgent in many countries." Similarly, Lee (1984) shows that UPE is unlikely to be achieved by 2020 unless unit costs are reduced, or a greater share of the GNP is devoted to primary schooling.

The Importance of Teachers' Salaries

Any attempt to reduce education costs is bound to focus on teachers' salaries initially, since they represent at least 70 percent of the total current costs of education in many developing countries and more than 90 percent of the current costs of primary education.

Various factors account for the high proportion of expenditure devoted to teachers' salaries. In many countries, teachers receive high salaries in relation to the general level of wages and salaries, but countries differ considerably in the ratio of teachers' salaries to GNP per capita or to average wages and salaries. In many African countries, for example, teachers' salaries are higher in relation to national income than in other developing countries. According to a recent survey of 59 developing countries, the average salary of primary-school teachers in Francophone Africa is 1.8 times the average in other developing countries (Eicher 1984). In Ethiopia, the average primary-school teacher's salary is ten times the GNP per capita. At the secondary level, the ratio of teachers' salaries to GNP is much higher. In Tanzania and Ethiopia, secondary-school teachers receive more than twenty times GNP per capita, which is very much higher than in developed countries. These findings indicate that teachers' pay scales are likely to exert a strong influence on the level of total expenditure.

Zymelman (1982b) has shown that substantial savings may be made in the medium term by marginally changing teachers' pay scales. This possibility has two implications: that teachers' pay scales are an important policy variable, even though the political power of teachers in many developing countries makes governments cautious about using it; and that the effect of pay scales has considerable bearing on the future costs of educational investment. The composition of the teaching force with respect to age and qualifications is a powerful determinant of total expenditure on teachers' salaries. One of the most frequent causes of underestimation of education costs is the failure to take into account the increased age or qualifications of teachers, particularly where pay scales provide age-related increments and are related to qualifications.

Coombs and Hallak (1972) emphasize that although teacher training costs are likely to be a modest fraction of total education costs, they can have a large multiplier effect on future teacher costs, not only by adding to the number of trained teachers, but by raising the level of teacher qualifications, and thus increasing average teacher salaries. A vivid example of this effect is provided in a case study of cost analysis in Uganda (Bennett 1972a), which examined a proposal to finance the building of four large regional teacher training colleges with the help of USAID. The proposed colleges would replace twenty-six small colleges and would contribute to a rapid improvement in the quality of the teaching force by training only grade 3 teachers. The effect would be to increase future costs per primary school pupil by 30 percent because of the salary differential between grade 2 and grade 3 teachers. Bennett has shown that this calculation was crucial in persuading the government of Uganda to revise the proposal since it indicated that the project was too costly, even if it was aid supported.

Another important determinant of total teacher salary costs is the average teacher-pupil ratio. Many countries are increasing teacher-pupil ratios and reducing average class size in the hope of improving educational quality. One review of studies in this area (Haddad 1978) has found no consistent evidence, however, of a relationship between class size and pupil achievement. In fact, a number of studies have shown that students perform better in larger classes, although there is some evidence that class size may not be a continuous variable; that is, within a certain range, small changes in class size would not affect pupil achievement. If, however, classes are simply grouped according to "large," "medium-small," and "very small," there is some evidence that small classes are significantly better than larger classes, although a change in average class size has no measurable effect within the same size grouping.

Even though it is impossible to prove a direct link between class size and pupil performance, class size has a considerable effect on costs because of the additional expenditure on teachers, classroom equipment, materials, and administration. In the absence of evidence that the additional cost generates additional educational benefits, it may be more cost-effective to allow a modest increase in class size and invest the annual savings in more teaching materials or textbooks. According to Farrell and Schiefelbein (1974), a 15 percent increase in the average class size in Chile would reduce the annual education budget by 5 percent; this saving could then be used to achieve significant quality improvements. In contrast, there is little scope for increasing the pupil-teacher ratio in countries such as those of West Africa where in 1978 the average class size was greater than 50 (Eicher 1984, p. 72). Sometimes, a low pupil-teacher ratio reflects a particular policy decision. In Mauritius, for example, the low pupil-teacher ratio is the result of the government's decision, for cultural and political reasons, to provide oriental language teaching for all primary school pupils. Although no general conclusions can be drawn about the feasibility of increasing average class size in developing countries, the fact that teachers' salaries exert such a powerful influence on total costs means that teachers' salaries must be carefully scrutinized in any attempt to reduce educational costs.

Treatment of Capital Costs

Because capital costs have great bearing on decisions about educational investment, special attention must be given to measuring and defining these costs. In the first place, it is essential to estimate capital costs accurately and to measure the annual cost of capital correctly, taking into account amortization and depreciation. Frequently the cost of projects is underestimated because of faulty assumptions or treatment of capital costs. Second, it must be recognized that capital costs entail future recurrent expenditure in the form of salaries for teachers to staff new schools, materials for these teachers, maintenance and repairs of buildings and equipment, and eventual replacement. All too often such items are neglected or underestimated, with the result that a capital project appears more profitable than it really is.

As noted earlier, capital expenditure is incurred to acquire goods and services that will be of use over a long period, whereas recurrent expenditure purchases goods and services of immediate, but short-lived usefulness. The problem in calculating the total cost of a project is how to add these two categories of expenditure and allow for the differences in time scale. In other words, we want to aggregate a *stock* of capital that is purchased at one point of time (but from which services are consumed over a period of time) and a *flow* of services that are consumed as they are produced.

This calculation is particularly important when the costs of introducing new media such as radio or television are being estimated, since a high proportion of the total costs is for the purchase of equipment that is expected to last for many years. If this equipment is assumed to have a useful life of ten years, then the initial capital cost can be divided by 10 to provide a measure of the annual *depreciation* of the asset. To assume that the annual cost of a project simply consists of annual current expenditure plus depreciation of capital, however, would be to seriously underestimate the social opportunity cost of a project. The purchase of a large piece of equipment or a building locks up resources for a number of years, with the result that alternative opportunities to invest these resources are forgone. The loss of the alternative returns, or interest, must be counted as an additional cost, and the capital must be *amortized* over its expected lifetime in such a way as to take account of the loss of interest as well as depreciation.

This can be done by *annualizing* capital costs, using a discount rate that represents the interest forgone (that is, the opportunity cost of capital). This annualization is sometimes called *imputed rent* since an alternative to purchasing the capital asset is to rent it. In the case of public investment, the opportunity cost is the *social discount rate*. Because of the problems of identifying the social discount rate, some cost calculations use alternative discount rates—say 7.5 percent and 10 percent—in annualizing capital costs. Others use shortcuts; for example, the original cost of the capital can simply be multiplied by the discount rate and this annual interest charge added to the annual depreciation. This is only a rough approximation, however, since it ignores the changing value of the capital asset over its lifetime (Jamison 1977).

The correct way to allow for both interest forgone and depreciation is to calculate an annual capital cost, or annualization factor, which is dependent on the social rate of discount, the lifetime of the capital, and the original cost of the capital.[4] In the terminology of discounted cash flow (see chapter 3), the annual capital cost is the present value of the annual sum required to repay the original cost of the capital over its assumed life.

From the value of annual capital cost calculated in table 7-2, it can be seen that the actual value of the annual capital cost depends critically on the assumed rate of interest and lifetime of the capital. If the social discount rate is 15 percent, for example, extending the lifetime of equip-

Table 7-2. *Annual Capital Cost of Investment of $1 Million at Varying Rates of Interest and Life of Equipment*
(thousands of U.S. dollars)

Life of equip-ment (years)	Rates of interest			
	7.5 percent	10 percent	12.5 percent	15 percent
5	247	264	281	298
6	213	230	247	264
7	189	205	223	240
8	171	188	204	223
9	157	174	191	210
10	146	163	180	199
11	137	154	172	191
12	129	147	165	184
13	123	141	160	179
14	118	136	155	175
15	113	131	151	171

Source: Wagner (1982).

ment costing $1 million from five to six years reduces the annual capital cost by $34,000 ($298,000 − $264,000); the same effect would be produced by reducing the discount rate from 15 to 10 percent, and assuming a life of five years.

In a discussion of the evaluation of new educational media, Wagner (1982) points out that this calculation makes a significant difference to the costs of a project. In this example it is almost 3.5 percent of the total capital cost. Yet the rate of interest and assumed life of the equipment are seldom discussed in detail in the assessment of the costs of a project, and more attention is paid to items of far less quantitative importance. This is a significant source of error in some calculations of capital costs, particularly those related to educational technology projects. Indeed, some cost estimates ignore the problem of interest altogether and use a zero discount rate. The use of an appropriate social discount rate is not just a theoretical nicety, but can make a significant practical difference in the assessment of the real costs of a project (Jamison 1977). A zero interest rate implies that the project planner is indifferent to the choice between spending a million dollars now or doing so ten years from now. Given the scarcity of funds for capital investment in developing countries, this position is obviously untenable, and to assume otherwise can lead to a serious underestimate of the costs of an instructional technology project, and an overestimate of its advantages compared with traditional systems, which involve much less capital expenditure.

One cost study of educational television in El Salvador has dismissed the problem on the grounds that "the inclusion of interest charges would not have made much practical difference . . . while opening a Pandora's Box of theoretical arguments, imputations and adjustments" (Speagle 1972, p. 228). A Pandora's box it may be, but the inclusion of interest charges does make a practical difference, as table 7-3 shows. In the case of El Salvador, costs are underestimated by 20 to 33 percent, depending on the rate of discount. Realistic assumptions about both the expected life of a capital asset and the social discount rate are particularly important in the assessment of investment projects that have a large capital component, such as radio or television. Capital costs account for only a small proportion of the total costs of traditional education, however. In general, the typical social costs of primary and secondary schooling in developing countries can be broken down as follows: teachers' salaries, 55 percent; other recurrent expenditure, 10 percent; capital cost, 5 percent; earnings forgone, 30 percent.

In the case of educational technology projects, however, the capital component is much greater and the capital equipment is expected to be used over a long period. This means that not only is it important to make realistic assumptions about interest rates and the life of the capital, but

Table 7-3. *The Extent of Cost Underestimation*
in Analyzing Ongoing Instructional Technology Projects
at Various Interest Rates (r)

Instructional technology project	Average cost per student (in 1972 U.S. dollars) at r =			Cost underestimate (in percent) if r = 0 is used and true r =	
	0 percent	*7.5 percent*	*15 percent*	*7.5 percent*	*15 percent*
Radio-based					
Nicaragua	3.65	3.86	4.07	5.4	10.3
Radioprimaria	12.63	13.12	13.72	3.8	8.0
Tarahumara	35.94	42.20	49.34	14.8	27.2
Thailand	0.29	0.35	0.41	17.1	29.3
Television-based					
El Salvador	19.72	24.35	29.37	19.0	32.9
Hagerstown	51.54	54.23	57.78	5.0	10.8
Korea, Rep. of	2.76	3.22	3.74	14.3	26.2
Stanford ITV	146.60	159.20	175.10	7.9	16.3
Telesecundaria	23.02	24.27	25.74	5.2	10.6

Source: Jamison, Klees, and Wells (1978), p. 34.

that the time scale may be crucial in determining whether a project is profitable or not. Jamison, Klees, and Wells (1978) have developed a methodology for examining the costs of educational technology that explicitly takes into account different time perspectives.

This step is necessary because the question of whether to introduce new media is quite different from the question of whether to continue or expand an existing project. The two decisions will depend on different analyses of costs. After a project has been introduced, the costs already incurred are *sunk costs* and should not be taken into account in assessing the profitability of continuing or extending the project in the future. What is needed is a method of calculating future costs on the basis of different time perspectives and different time horizons for the life of the project. For example, a planner may want to answer a series of questions: What would be the costs of introducing educational television if it is assumed the equipment will last for twenty years? What would the average costs be if the project is abandoned after only ten years? What would be the costs of continuing the project after it has been running for five years? All these questions require a different estimate of costs because, as we have already emphasized, the planner is not indifferent to whether costs are incurred now or in the future.

The method advocated by Jamison, Klees, and Wells consists of calculating average costs (AC) as seen from any year (i) of a project, with a time horizon, or expected life of the project, (j). AC_{ij} can be calculated[5] on the basis of different values for both i and j, discounting both future costs and future utilization, in order to estimate the present value in year (i) of the costs of educating N students up to year (j). Obviously, such a calculation depends on the rate of discount adopted, which in turn represents the social time preference rate or the social opportunity cost of capital. Jamison, Klees, and Wells (1978) suggest using different discount rates, to correspond to different assumptions about time preference.

This means that the average cost of a project depends on the planner's time perspective. If costs have already been incurred and the planner is looking ahead to future costs, then "bygones are bygones" and past costs should not be taken into account in estimating future costs. When the project is completed, however, the planner may wish to calculate the average cost per student over the whole life of the project; in that case, it would be necessary to include all capital costs, but on an annualized basis. Thus, the question "What is the average cost per student?" depends on the time scale of the project and the time at which the analysis takes place in relation to the total life of the project. In general, "the further into the project we are, the less expensive it is to continue, due primarily to many initial project development expenses becoming sunk costs" (Jamison, Klees, and Wells 1978, p. 225). It also means that the higher the opportunity cost of capital, the higher the cost of the project.

Because this approach seems to introduce an element of uncertainty into cost calculations, it is frequently misunderstood and arouses considerable controversy. A recent discussion of the costs of new educational media (Wagner 1982) uses the concept of AC_{ij}, for example, but argues that it is necessary only to discount future costs, not student utilization. The rationale for discounting both costs and student numbers is that planners are not indifferent to whether costs are incurred now or in the future, nor are they indifferent to whether students are educated now or in the future. This remains a controversial issue, however, and by no means all cost analyses of educational technology projects use the methodology recommended here.

In fact, even the annualization of capital costs discussed earlier is rejected by some authors. Eicher (1977), for one, argues that although it is correct from the wider economic standpoint to include interest forgone in amortization calculations, it is not necessary if the purpose of the cost analysis is simply to compare the relative costs of different educational methods. If the alternatives have a different time pattern of costs, however, then this is a factor to be taken into account.

Evaluations of World Bank projects do try to take into account the social opportunity cost of capital and the time preference of the government, and it is recognized that both may vary among countries, depending on individual circumstances. It is also recognized that the choice of whether to launch a project differs from the choice of whether to continue or expand an existing project with respect to the costs that have already been incurred and the timing of future costs and future benefits. The example given below shows how cost-benefit analysis techniques that take into account the timing as well as the magnitude of future costs and savings can be used to evaluate the cost savings expected from new educational technology.

Before leaving the subject of capital costs, however, we must look at another potential source of underestimation of costs, namely, the underestimation of future recurrent costs that will be necessary to operate and maintain a project after the capital expenditure has been incurred. When a project is first assessed, it is all too easy to concentrate on the costs in the early years and to underestimate the future costs of repairs, maintenance, and support services. If the project is to be financed through international or bilateral aid, there is a particular danger that the future recurrent cost implications may be underestimated because planners are concentrating on initial capital costs.

Yet any capital expenditure will involve future recurrent expenditure, which in some instances may be quite substantial. The installation of new capital equipment will obviously give rise to future spending on repairs, maintenance, and renewal, but may also require expenditure on staff training and supervision and on the production of support materials. The building of a new school will require additional teachers and administrative staff, materials, and equipment and will also introduce operational costs for items such as heating, lighting, and maintenance. A new teacher training college may be financed through international aid, but the cost of the teachers' salaries, when the trained teachers are eventually employed, will fall on the government. The fact that the future recurrent cost implications of capital projects have so often been underestimated in the past has led governments in developing countries to be increasingly cautious about financing capital expenditure for educational investment through loans, since the loan repayments may fall due just at the time that the underestimation of recurrent costs is becoming apparent. Similarly, the fear that future recurrent costs may be higher than anticipated may make governments reluctant to accept aid for capital projects, as shown by the example of teacher training in Uganda, discussed above.

An example of the underestimation of future recurrent costs can be found in the Ivory Coast, where an ambitious program of primary education by television was introduced in 1971. The project was partly financed through aid and loans from Unesco and the World Bank, and it has been the subject of various cost analyses (IIEP 1972; Eicher and Orivel 1980). The original calculation of the costs of installing television sets in schools ignored depreciation, and thus future recurrent costs were underestimated by at least 30 percent (Chau 1972).The recurrent costs of electricity supply and batteries were also underestimated—the costs of maintenance and electricity each year are higher than the purchase price of a television set (Eicher and Orivel 1980). The cost of producing support materials, particularly the cost of paper for written materials, also proved a heavy burden. As a result, the total costs proved higher than anticipated, partly because of the failure to estimate future recurrent costs accurately, and partly because of unforeseen inflation.

In some projects it is possible to calculate future recurrent costs as a percentage of initial capital costs by means of a formula. In the case of education projects, however, differences between projects are so significant that attempts to use such a formula are unlikely to be reliable. Jamison, Klees, and Wells (1978, p. 61) examine various reasons for the underestimation of recurrent costs, including the possibility of built-in bias in the original assessments: "It is often politically desirable for factions within a Ministry of Education, who desire a particular project, to underestimate the project's costs." The problem is not confined to education; analyses of capital-intensive public sector projects in other areas, including defense, regularly find that subsequent costs exceed original estimates by a factor of 2, 3, or 4. The fact is that formulas for cost projections or for calculating the recurrent cost implications of capital projects, like other techniques of cost analysis, cannot replace careful human judgment. Jamison, Klees, and Wells (pp. 59–60) conclude that cost analysis and cost projections are an art, rather than a science, and require ingenuity as well as careful judgment.

Careful judgment is needed to identify and forecast future recurrent expenditure likely to arise as a result of capital investment. Such expenditure would include teacher costs and allowance for any necessary retraining or upgrading that may increase average salaries; other staff costs; support materials; maintenance and repairs; operating costs (electricity or other power); and depreciation. When such a list has been compiled, it may suggest that a capital-intensive project, even if financed through foreign or international aid or low-interest loans, will impose too heavy a burden on public funds. In such a case, a government may decide to reject the project, or it may decide to explore the feasibility of using cost-recovery mechanisms (see chapter 6). Still another option would be to search for ways of reducing costs, for example, by analyzing patterns of unit cost and the determinants of cost.

The Pattern of Unit Costs of Education

Comparisons between unit costs of education in developed and developing countries during the 1970s reveal some striking differences. Unit costs at all levels were much higher in developed than in developing countries, and the gap between the poorest and the richest countries has been growing. If unit costs are compared with GNP per capita, however, then education represents a much heavier economic burden in developing countries, particularly at the secondary and higher levels. On the average, the cost per student in higher education in developed countries was 55 percent of GNP per capita in the early 1970s, whereas developing countries on the average spent five times the level of GNP per capita, and many African countries spent more than ten times the average per capita income on each student (Zymelman 1976).

Although these heavy burdens have been reduced somewhat (see table 7-4), unit costs at the secondary level still represent a much heavier burden (with respect to GNP per capita) in Africa than in developed countries or in Asia or Latin America. Unit costs of higher education are half the annual GNP per capita in OECD countries, one-and-a-half times in the Middle East and North Africa, and eight or nine times that in West and East Africa.

Another way of looking at this striking difference is that each university student costs two or three times as much as one primary school pupil in developed countries, but costs ten times as much in most developing countries, and thirty or forty times as much in Africa. These differences raise a number of questions about the possibility of cost reductions: Are these enormous cost differences due to inefficiencies, waste, and extravagant provision of facilities, as has sometimes been suggested, or are they due to the low levels of enrollment in higher education in developing countries? In other words, is it possible for developing countries, particularly in Africa, to exploit economies of scale in higher education?

The fact that there is considerable evidence of economies of scale in higher education (Psacharopoulos 1982; Lee forthcoming) suggests that one reason for the very high unit costs in Africa is the low level of enrollment. Thus, expansion of higher education may enable some developing countries to reduce cost per student. The level of enrollment is not the only explanation for these high costs, however, or for the striking differences in unit costs by level in developing countries.

Since we have already emphasized the high proportion of recurrent costs devoted to teachers' salaries, the first question to ask is whether the relatively high costs in developing countries are due to low student-staff

Table 7-4. *Unit Cost/GNP Per Capita and Enrollment Ratios for Primary, Secondary, and Higher Education, by Region*
(percent)

Region	Primary				Secondary				Higher			
	Unit cost/ GNP per capita		Enroll- ment ratio		Unit cost/ GNP per capita		Enroll- ment ratio		Unit cost/ GNP per capita		Enroll- ment ratio	
	1970 -72	1976 -79	1970 -72	1976 -79	1970 -72	1976 -79	1970 -72	1976 -79	1970 -72	1976 -79	1970 -72	1976 -79
East Africa	20	19	61	73	124	68	7	13.6	927	907	0.47	0.67
West Africa	24	25	43	43	142	102	6.6	9	1,405	838	0.36	0.73
East Asia and Pacific	12	11	94	92	24	19	28	39	179	118	2.54	6.00
South Asia	9.6	8	46	67	34	16	18	17	260	119	1.61	2.58
Europe, Middle East and North Africa	15	13	74	81	47	26	25	34	306	151	2.10	4.00
Latin America and the Caribbean	11	9.3	93	93	22	15	26	38	121	88	7.90	6.00
"Average" developing country	15.3	14	68	75	65.5	41	18.4	25	533	370	2.50	3.30
OECD average	16	22	98	100	21	24	80	80	55	49	14.60	21.00

Source: Zymelman (1982a), p. 49.

186

ratios. An analysis of data for higher education in 123 developing countries and 20 developed countries (Lee forthcoming) shows little difference between developed and developing countries in this regard. There are considerable differences in pupil-teacher ratios at the primary level, however. On the average, the primary school pupil-teacher ratio is 36 in developing countries and about 20 in developed countries. Data for individual countries show even greater differences between levels. In some African countries, the average pupil-teacher ratio in primary schools is greater than 50, while the student-staff ratio in universities is often more favorable than in developed countries.

Another striking difference between the two levels is that in many developing countries students in higher education pay no fees and even receive free books and free board and lodging, whereas primary school pupils must pay fees as well as all living expenses (see chapter 6). These generous subsidies of students in higher education are, of course, reflected in the high unit costs at this level.

One explanation for high unit costs is the cost of equipment and materials, particularly in laboratory courses, and the high costs of buildings. These factors are reflected in the difference between the unit costs of arts and science. In Asia, for example, the total capital cost per university place in arts subjects is fourteen times higher than a primary school place, but in science it is forty-five times higher (table 7-5).

These differences in costs per student by level of education and subject are the main reason for the marked differences in rates of return (see chapter 3), since the difference in costs is not matched by a similar difference in the lifetime earnings of graduates. Does this mean that a planner faced with a choice between investing in primary education and higher education or between arts and science courses, must always choose the lower-cost option? In view of the significant benefits from higher education, this conclusion would not be justified. The alternative is to try to reduce unit costs in higher education or in expensive courses in order to make the cost-benefit ratio more favorable.

Cost analysis has been used by a number of individual institutions in an effort to identify possible cost reductions. A study of Makerere University, for example, found that it would be possible to reduce unit costs by more intensive use of buildings and by reductions in staff-student ratios (Bennett 1972b). Although such studies do not provide general rules that could be applied throughout developing countries, they do suggest that detailed cost analysis at the institutional level may be useful in identifying the scope for cost reduction. For one thing, detailed studies may be helpful in developing cost functions, which can then lead to a better understanding of the relationship between fixed and variable costs and between average and marginal costs.

Table 7-5. *Capital Cost per Student Place in Selected Asian Countries*
(1964 U.S. dollars)

Educational level	Area per student place (square meters)	Build- ing cost per place	Equip- ment cost per place	Total capital cost per place
Primary	1.3	35	11	46
Lower secondary				
General	3.0	120	66	186
Vocational	4.0	160	101	261
Upper secondary				
General	3.0	120	66	186
Technical	5.0	200	146	346
University				
Science and related	16.0	960	1,120	2,080
Arts and related	6.0	360	300	660

Note: Figures are averages for Afghanistan, Laos, Nepal, Burma, Cambodia, India, Indonesia, Iran, Mongolia, Pakistan, Viet Nam, Sri Lanka, Taiwan, Rep. of Korea, Malaysia, Philippines, Singapore, and Thailand.
Source: Psacharopoulos (1980), p. 87.

The Relationship between Average and Marginal Costs: Cost Functions

We argued above that decisions about whether to expand existing systems or institutions should be based on marginal rather than average costs. The relationship between average and marginal costs is determined by the cost function, which shows how total costs change in relation to the number of units produced (that is, the number of students or pupils being educated). A very simple cost function might be $TC = \$100,000 + 200N$, which means that costs can be divided into a fixed element ($FC = \$100,000$) that is invariant with student numbers (N), and a variable element ($VC = \$200$).[6] In this case, the marginal cost (MC) of one additional student is equal to the variable cost of $200, and average cost ($AC$) will fall with every increase in N, as the fixed cost is shared between a larger number of students. Eventually, however, as N becomes very large, the fixed cost per student will become negligible, and AC will approximate MC ($200).

This is what is meant by the term economies of scale. If marginal cost is lower than average cost, any increase in student numbers will lower the average cost per student. If marginal cost equals average cost, then there are constant returns to scale, and average cost will remain the same, regardless of student numbers. If, however, the marginal cost of educating an additional student is higher than the average cost, an increase in numbers will raise average costs. This situation is referred to as diseconomies of scale.

The question is whether there are economies of scale in education. The answer depends on the shape of the cost function and the relationship between fixed and variable costs. In the case of new educational technologies, the fixed and variable costs are fairly obvious. Production and transmission costs of radio or television, for example, do not vary directly with the number of students, whereas reception costs are much more directly influenced by numbers. Thus, economies of scale in educational broadcasting will be significant since marginal cost will be lower than average cost. Increasing the size of the audience will therefore cause average cost to fall.

Preliminary evaluation of the cost of educational television in the Ivory Coast (Chau 1972) has suggested that average cost would vary with the number of pupils receiving television instruction (see figure 7-1). Average costs of production, transmission, and reproduction fall as the number of pupils increases, but the fall is most marked in the case of production costs, since these are entirely fixed costs. This forecast was made before television was introduced in the Ivory Coast, but Eicher and Orivel (1980) were able to use actual experience to estimate a cost function. According to their calculations, average cost falls sharply until the number of pupils receiving television instruction reaches about 300,000, after which economies of scale would be exhausted, and average cost would be almost equal to marginal cost (figure 7-2). In the case of a capital-intensive project such as educational television, there is obvious scope for economies of scale, since a high proportion of total costs are fixed costs associated with establishing a broadcasting network and producing programs.

Cost functions for educational technology have been analyzed extensively in case studies (Spain, Jamison, and McAnany 1977) and reviews (Jamison and McAnany 1978; Eicher et al. 1982; Perraton 1982). Many of the case studies employ cost-effectiveness analysis. That is, they compare the costs of alternative methods of teaching on the assumption that effectiveness, in terms of the quality of instruction, is equal. The most cost-effective method—that is, the method that educates students to a given standard at least cost—can then be identified. (Effectiveness is discussed in chapter 8.)

That unit costs vary greatly among projects can be seen from the hourly production costs in twelve educational television projects listed in table 7-6 and in summaries of the costs of fourteen distance-teaching projects

Figure 7-1. *Forecast of Number of Pupils Receiving Instruction
by Television and Average Cost per Pupil, the Ivory Coast*

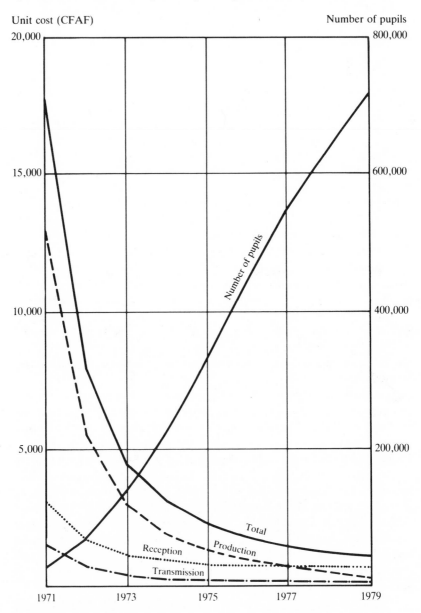

Source: Chau (1972).

Figure 7-2. *Average Cost Function for Educational Television,
the Ivory Coast, 1975*

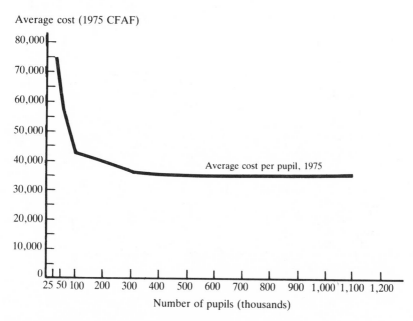

Average cost (1975 CFAF)

Average cost per pupil, 1975

Number of pupils (thousands)

Source: Eicher and Orivel (1980), p. 137.

involving radio and television in developing countries (Perraton 1982).
Differences in the level and pattern of costs are illustrated in table 7-7,
which shows a sample of the cost functions in different projects. The
general experience, however, is that the proportion of fixed costs in such
projects is very high, often exceeding 50 percent, whereas in traditional
education fixed costs are often much lower than variable costs. This
means that enrollment levels are particularly important in such projects
since average costs will fall rapidly, as enrollment increases, until a
minimum threshold is reached. The cost function illustrated in figure 7-2
suggests that in the Ivory Coast the minimum threshold in 1975 was about
300,000 pupils. If the number of pupils had been less, it would have been
possible to reduce average costs by expanding enrollment. After the
threshold had been reached, economies of scale would have been exhausted, fixed costs would have been negligible, and both marginal and
average costs would have been roughly equal to variable costs.

It is impossible to generalize about the minimum or the optimal level of
enrollment in educational technology projects, however, because this
level will depend on the level of education, the proportion of fixed to

Table 7-6. *Hourly Production Costs of Televised Programs*
in Different Educational Television Projects

Project	Annual number of hours produced	Average cost of each hour produced (1980 U.S. dollars)
Ceará (Brazil)	300	2,750
El Salvador	333	5,665
Hagerstown (United States)	1,440	1,450
Ivory Coast (in school)	201	25,900
Ivory Coast (out of school)	17	51,200
Korea, Rep. of	n.a.	3,220
Maranhão (Brazil)	525	1,815
Open University (United Kingdom)	288	18,150
Senegal (TSS)	49	16,600
Stanford (United States)	6,290	175
Telecurso (Brazil)	75	53,800
Telesecundaria (Mexico)	1,080	925

n.a. Not available.
Source: Eicher et al. (1982), p. 56.

variable costs, the choice of media (television has higher average costs than radio, for example), and a number of other factors that will vary between countries. It might be asked, for example, what proportion of the intended audience may be expected to own television or radio sets? What are the costs of electricity and maintenance? The answers to these questions will vary but will help to determine the extent of economies of scale in projects involving new educational media.

The concepts of cost functions, fixed and variable costs, and average and marginal costs are clearly relevant to education technology projects, even though it may be difficult to measure them precisely. The existence of economies of scale in such projects has been amply demonstrated, and there are many examples of educational cost functions calculated for new technology projects in both developed and developing countries. Ideally, the decision whether to introduce new technology should be preceded by a careful analysis of cost functions, although in practice such decisions often represent a political act of faith, rather than a careful analysis of cost functions (Unesco 1980).

The question of whether there are economies of scale in traditional institutions is more difficult. This depends on whether there is spare capacity, the existence of which would mean that marginal cost is lower than average cost. Economies of scale have been investigated in the

United States (Riew 1966; Cohn 1968; Maynard 1971) and other developed countries, but only a few studies have been done in developing countries. Among these are recent studies based on international comparisons of average cost per student and enrollment in higher education (Psacharopoulos 1982; Lee forthcoming), both of which demonstrate that economies of scale do exist in developing countries. Psacharopoulos compared average cost and enrollment data for eighty-three countries and concluded on the basis of the pattern illustrated in figure 7-3 that average costs are lower in countries with high levels of enrollment and that the cost per student with respect to per capita income declines sharply after an enrollment ratio of 2 to 3 percent and steadies out thereafter. This pattern corresponds to the level of enrollment in countries such as Zambia, Congo, Pakistan, Nepal, Mauritius, Morocco, El Salvador, and Cambodia, and means that average cost in such countries may be expected to fall as university enrollment increases.

Lee (forthcoming) has further examined cross-country evidence of economies of scale in higher education and concluded that average costs decline sharply until an enrollment of about 500 students is reached; significant savings in terms of lower average costs can still be achieved up to an enrollment level of about 10,000 students, but after this, further expansion will lead to only a minimal fall in average cost.

This finding—namely, that small institutions are associated with high average costs—has practical implications for the design of higher-education projects. Cost factors are not the only consideration in determining the size of institutions, of course, since geographical and social factors, as well as economic ones, all affect the final choice. The point is, however, that because there are significant economies of scale in education, the planner should take into account marginal costs as well as average costs.

As noted earlier, marginal cost calculations were used in planning the expansion of the University of Makerere between 1966 and 1971. Bennett (1972b) subsequently showed that although the original capital cost, per place, of existing facilities in Makerere was estimated to be 100,000 shillings, an expansion had recently taken place at a cost of only 35,000 shillings per place, and it was estimated that further expansion could take place at even lower cost per student, if facilities were used more intensively. Thus, considerable spare capacity was present at the time of the cost analysis. This is the clue that tells us whether marginal costs are below or above average costs. If there is spare capacity—for example, in the form of underutilized buildings, equipment, or teaching staff who could devote more time to teaching—then marginal costs will be below average costs since additional students could be accommodated without incurring expenditure for extra buildings, equipment, or staff. To identify

Table 7-7. Costs of a Sample of Some In-School Distance-Teaching Projects

Project	Units for measuring N[a]	Value of N at time of study	Cost function (1978 U.S. dollars at 7.5 percent discount rate)	Average cost for given N (1978 U.S. dollars)
Brazil, Minerva Madureza Project	Number of subject equivalents a year	1977 $N = 177,000$	$TC(N) = 907,120 + 19.7N$	24.8
Brazil, Bahia Madureza Project	Number of subject equivalents a year	1977 $N = 8,000$	$TC(N) = 437,900 + 12.6N$	67.4
Malawi Correspondence College	Number of students reached per year	1978 $N = 2,800$	$TC(N) = 132,000 + 117N$	160.0
Mauritius College of the Air	Number of course enrollments	1976 $N = 12,000$	$TC(N) = 143,800 + 1.98N$	14.0

Brazil, Maranhão FMTVE	Number of students reached per year	1976 $N = 13{,}000$	$TC(N) = 1{,}551{,}000 + 116N$	235.3
Brazil, Cerá Educational Television Project	Number of students reached per year	1978 $N = 19{,}800$	$TC(N) = 971{,}000 + 87N$	136.6
Mexico, Radioprimaria	Number of students reached per year	1972 $N = 2{,}800$	$TC(N) = 57{,}500 + 139.2N$	160.0
Mexico, Telesecundaria	Number of students reached per year	1972 $N = 29{,}000$	$TC(N) = 842{,}700 + 204.1N$	235.7

a. N is the measure of the degree of utilization of the system; a subject equivalent is defined as the enrollment of a student for a course in one subject for one semester.

Source: Perraton (1982), pp. 258–59.

Figure 7-3. *Average Cost per Student and Level of Enrollment*

Average cost per student (U.S. dollars)

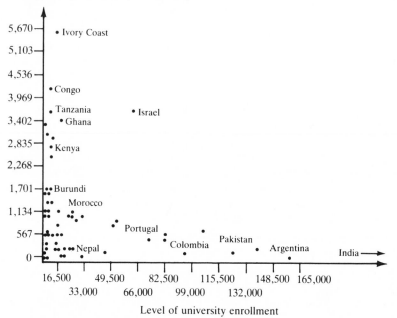

Level of university enrollment

Source: Psacharopoulos (1980).

spare capacity or underutilized resources, it is necessary to examine particular situations or institutions. Thus, no general formula can be used to determine the relationship between average and marginal costs and to establish whether there are economies of scale. Instead, cost functions must be analyzed to show how total costs vary with respect to student numbers, or else resources must be carefully scrutinized, to test whether they are fully utilized.

Assessments of World Bank education projects at the time of project completion show that several have underutilized facilities, that is to say, spare capacity. For example, one recent analysis noticed consistent underenrollment in nonuniversity technical training projects. Since many of these underutilized institutions have more costly equipment and higher teacher-student ratios than general institutions, their spare capacity combined with high unit costs indicates extreme inefficiency, given the constraints on public expenditure. When there is evidence of spare capacity, expansion of technical training in existing institutions should lead to a reduction in unit costs.

Economies of scale have also been identified in the diversified secondary schools that have been built through World Bank assistance to

Colombia. The decision to build nineteen comprehensive secondary school, (Institutos Nacionales de Educacion Media, or INEMs) was based on the belief that the large size of each school would promote economies of scale. Between 1975 and 1979, average cost per student fell from 8,058 to 5,166 pesos, but there were considerable differences between individual schools; the fact that the average cost in some small schools was 60 percent higher than in larger schools suggests that an expansion of schools with high average costs might enable them to become more efficient (Psacharopoulos and Loxley forthcoming).

This discussion, like most of the analysis of cost functions in education, has been concerned with the effects of expansion. Some countries, however—including some developing countries—face a decline in the primary age group in the near future. The assumption is often made in economic textbooks that the marginal savings that can be expected from a small decrease in numbers can be calculated in exactly the same way as the marginal increase in costs due to expansion. That assumption must be handled with caution since there are few examples of marginal savings achieved due to a contraction of educational systems or institutions, although a number of developed countries have experienced falling rolls as a result of a falling birth rate. Some developing countries will experience this fall in the next decade, as family planning policies begin to show results. It is not at all obvious, however, that costs that were considered variable when student numbers were increasing will still be considered variable when numbers are declining. In particular, the question of whether, and how quickly, the number of teachers can be adjusted in response to falling numbers depends on the conditions governing teachers' employment. If teachers enjoy security of tenure or sufficient political power to fight compulsory redundancies, then it may be difficult to reduce the number of teachers employed, and a reduction in student numbers may lead to an increase in teacher-student ratios and an increase in average costs. Thus, it may take considerable time before a reduction in student numbers is translated into a reduction in costs, and marginal savings may not be apparent for a long period. Analysis of the relationship between average and marginal costs in a period of declining numbers is only just beginning in a few countries.

The Evaluation of Cost Savings in a Cost-Benefit Framework

Evidence from cost studies of distance teaching suggests that the use of educational media, particularly radio, may be less costly in terms of average cost per student than conventional teaching. To achieve these cost savings, however, it is necessary to invest heavily in capital equipment. To evaluate the profitability of such an investment in capital-intensive projects, it is necessary to find some way of comparing costs and benefits, or effectiveness. Before we look at that problem (see chapter 8), we should examine an attempt to carry out an economic evaluation of the cost-saving potential of educational technology.

A full cost-benefit analysis using the methodology described in chapter 3 would require data on the lifetime earnings of workers educated by different methods, and no such analysis has been conducted in a developing country. It is still possible, however, to evaluate potential cost savings in a cost-benefit framework by measuring monetary benefits in terms of the reduction in expenditure and by comparing this with the cost of investing in new technology. Such an approach was used, for example, in evaluating the investment in educational television in the Ivory Coast (Sirken 1979, 1982). The assumption was that primary school pupils could be educated with the aid of television in 5.5 years instead of the 8 years previously required because of dropout and repetition. A reduction in unit costs was therefore assumed and was multiplied by the total number of pupils educated through television to give an estimate of total cost savings for each year of the project (see table 7-8). Further savings were expected from using television to train teachers more efficiently.

The estimates of cost savings (table 7-8) were compared with the additional costs of educational television. By discounting both costs and benefits at 8 and 12 percent, we arrive at the figures shown in table 7-9. Clearly, if the opportunity cost of capital in the Ivory Coast is 8 or 12

Table 7-8. *Benefits of Television Project in the Ivory Coast*

Year	Savings per pupil (CFAF)	Pupils covered by project (thousands)	Total cost savings (millions of CFAF)		
			Pupils	Teacher training	Total[a]
1970–71	3,964	21	83	118	201
1971–72	3,964	63	250	354	604
1972–73	3,964	126	500	707	1,207
1973–74	3,964	210	832	1,061	1,894
1974–75	3,964	315	1,249	1,415	2,664
1975–76	3,964	430	1,705	1,707	3,412
1976–77	3,964	534	2,117	1,820	3,936
1977–78	3,964	627	2,485	1,752	4,237
1978–79	3,964	709	2,811	1,566	4,377
1979–80	3,964	720	2,854	1,044	3,898

a. Totals may not add up because of rounding.
Source: Sirken (1982), pp. 94, 96.

Table 7-9. *Benefit-Cost Ratio, Television Project, Ivory Coast*
(millions of CFAF)

	Costs			Benefits		
Year	Total costs	Present value at 8 percent	Present value at 12 percent	Total benefits	Present value at 8 percent	Present value at 12 percent
1968–69	568	526	507	0	0	0
1969–70	921	789	734	0	0	0
1970–71	694	551	494	201	160	143
1971–72	753	553	479	604	444	384
1972–73	839	571	476	1,207	822	684
1973–74	969	610	491	1,894	1,193	960
1974–75	1,120	653	506	2,664	1,553	1,204
1975–76	1,141	616	461	3,412	1,842	1,378
1976–77	1,177	588	425	3,936	1,968	1,421
1977–78	1,158	536	373	4,237	1,962	1,364
1978–79	988	424	284	4,377	1,878	1,256
1979–80	1,069	424	275	3,898	1,548	1,002
Total		6,841	5,505		13,370	9,796

Note: Benefit-cost ratio at 8 percent = 13,370/6,841 = 1.95. Benefit-cost ratio at 12 percent = 9,796/5,505 = 1.78.
Source: Sirken (1982), p. 97.

percent, then investment in educational television would be very profitable according to these assumptions. In fact, if these cost savings could be achieved, the rate of return on the investment would be 35 percent. This estimate is based on the assumptions about the extent of cost reductions in primary schools and teacher training. If, however, dropout and repetition rates did not fall as rapidly as was hoped, the savings would be smaller and the investment in television would appear much less profitable.

This is an example of how sensitivity analysis can be used to explore the implications of alternative assumptions about costs and benefits. If the number of pupils receiving television instruction increased, for example, the profitability of the investment would increase, but if the original cost estimates proved too low, the investment would appear less profitable. Sensitivity analysis is one way of allowing for the risks involved in educational investments and incorporating uncertainties into a comparison of costs and benefits.

Even if it is based on crude assumptions about cost, this form of analysis can provide a useful comparison of the cost savings and expected benefits of an investment project. Similar techniques have been used to compare the costs of alternative methods of vocational training, which differ in the timing of costs as well as their magnitude (McMeekin and Gittinger 1984). The value of this kind of systematic analysis is that it focuses attention on the need to reduce costs.

The Need for Cost Reductions

The recurring theme of this chapter and of many recent policy statements in developing countries is the urgent need for cost reduction in education. Data on trends in total costs and the recent slowing down of the growth in educational expenditure, coupled with continuing social demand, all indicate that the cost of educational investment must be reduced if its profitability is to increase, and if further expansion or improvement is to take place, despite budgetary constraints. This becomes all too obvious when we examine the recurrent and capital cost implications of the educational targets and reforms envisaged in a typical developing country (Hultin 1975). Typical objectives in the educational plans of most developing countries include expanded enrollments, quality improvements such as reduced student-teacher ratios, an increase in the proportion of qualified teachers, and an increase in the supply of materials, books, and the like. Each of these proposals may not be too costly if taken separately, but a combination could be extremely expensive.

Hultin estimated the separate and cumulative effects on recurrent costs of meeting such quantitative and qualitative targets at the secondary level. In some cases, costs would increase two or three times, and the cumulative effect of all the proposals would be that costs would increase eightfold between 1974 and 1983. Similar estimates for primary and higher education suggest that primary school costs would increase fivefold and higher-education costs fourfold if all the targets were achieved.

Such an increase would mean that education spending would rise from 4 to 14 percent of GNP, a rate quite outside the boundaries of feasibility. Lest it be assumed that this conclusion is due to the choice of excessive targets, Hultin emphasizes that the assumed targets are comparatively modest by the standards of developed countries and are commonly advocated in developing countries. His analysis of the cost implications shows, however, that these targets, while politically attractive, are seldom viable in financial terms. The conclusion is that few developing countries can afford to expand or improve their education systems without reducing costs. But what is the scope for cost reduction?

This chapter has shown that there are no panaceas, but that a number of potential cost savings reside in economies of scale and in the use of low-cost media, such as radio, which can reduce unit costs without imposing substantial capital burdens. Analysis of the determinants of costs shows the added importance of teacher salaries, however. Policies designed to increase the qualifications of teachers, raise the real level of teachers' salaries, or increase the teacher-pupil ratio can push costs up sharply. At the same time, policies designed to reduce teacher-pupil ratios or lower teachers' salaries may meet with considerable political opposition.

Our final conclusion, therefore, is that greater cost consciousness is needed in education. This means that we must not only analyze the behavior and determinants of costs, examine the utilization of resources to identify spare capacity, and question the traditional practices to identify possible cost reductions, but we must also link research on costs with research on effectiveness. This linking is the subject of chapter 8.

Notes to Chapter 7

1. The arithmetic average or mean is calculated from the formula

$$\bar{X} = \frac{X_1 + X_2 + \ldots X_n}{N} = \sum \frac{X}{N},$$

whereas the geometric mean is calculated from the formula

$$X_{GM} = \sqrt[N]{X_1 \cdot X_2 \ldots X_n},$$

or in other words,

$$\log X_{GM} = \frac{1}{N} (\log X_1 + \log X_2 \ldots \log X_n)$$

$$= \sum \frac{\log X}{N}.$$

2. Average cost $(AC) = TC/N$, where TC is the total cost of educating N students.

3. Marginal cost $(MC) = TC(N) - TC(N - 1)$.

4. The formula is

annual capital cost or annualization factor $(a) = \dfrac{C[r(1 + r)^n]}{[(1 + r)^n - 1]}$,

where r = rate of interest, C = original cost of capital, and n = assumed life of the capital. A more technical discussion of this formula is given in Jamison (1977) and Jamison, Klees, and Wells (1978).

5. The formula is

$$AC_{ij} = \frac{\sum\limits_{k=i}^{j} C_k/(1+r)^{k-i}}{\sum\limits_{k=i}^{j} N_k/(1+r)^{k-i}}.$$

More detailed discussion is provided in Jamison, Klees, and Wells (1978).

6. This is a simple linear function, which means that TC will increase in a straight line with every increase in N and MC is constant. The cost function could take many other forms, including a quadratic function or a hyperbolic function. For detailed discussion of different forms of cost function see Riew (1966), Cohn (1968), Maynard (1971), Psacharopoulos (1982), and Lee (forthcoming).

References

Bennett, Nicholas. 1972a. Educational Cost Evaluation: Uganda. In *Educational Cost Analysis in Action: Case Studies for Planners*. Vol. 3. Paris: International Institute for Educational Planning.

———. 1972b. The Use of Cost Evaluation in the Planning of Makerere University College. In *Educational Cost Analysis in Action: Case Studies for Planners*. Vol. 3. Paris: International Institute for Educational Planning.

Chau, Ta Ngoc. 1972. The Cost of Introducing a Reform in Primary Education. In *Educational Cost Analysis in Action: Case Studies for Planners*. Vol. 2. Paris: International Institute for Educational Planning.

Cohn, E. 1968. Economies of Scale in Iowa High School Operations. *Journal of Human Resources* 3, no. 4 (Fall):422–34.

Coombs, P., and J. Hallak. 1972. *Managing Educational Costs*. New York: Oxford University Press.

Eicher, J. C. 1977. Cost-Effectiveness Studies Applied to the Use of New Educational Media. Methodological and Critical Introduction. In *The Economics of New Educational Media*. Vol. 1. Paris: Unesco.

———. 1984. *Educational Costing and Financing in Developing Countries: Focus on Sub-Saharan Africa*. World Bank Staff Working Paper no. 655. Washington, D.C.

Eicher, J. C., and F. Orivel. 1980. Cost Analysis of Primary Education by Television in the Ivory Coast. In *The Economics of New Educational Media*. Vol. 2. Paris: Unesco.

Eicher, J. C., D. Hawkridge, E. McAnany, F. Mariet, and F. Orivel, eds. 1982. *The Economics of New Educational Media*. Vol. 3. *Cost and Effectiveness: Overview and Synthesis*. Paris: Unesco.

Farrell, J. P., and E. Schiefelbein. 1974. Expanding the Scope of Educational Planning: The Experience of Chile. *Interchange* 5, no. 2:18–30.

Fisher, G. 1971. *Cost Considerations in Systems Analysis*. Santa Monica, Calif.: Rand.

Haddad, Wadi D. 1978. *Educational Effects of Class Size*. World Bank Staff Working Paper no. 280. Washington, D.C.

Hultin, Mats. 1975. The Costs of More and Better Education in Developing Countries. In *Costing and Financing Education in LDC's: Current Issues*, ed. Mats Hultin and Jean-Pierre Jallade. World Bank Staff Working Paper no. 216. Washington, D.C.

International Institute for Educational Planning (IIEP). 1972. *Educational Cost Analysis in Action: Case Studies for Planners*. 3 vols. Paris.

Jamison, Dean. 1977. *Cost Factors in Planning Educational Technology Systems*. Paris: Unesco/IIEP.

Jamison, Dean T., and Emile G. McAnany. 1978. *Radio for Education and Development*. Beverly Hills and London: Sage.

Jamison, Dean, Steven Klees, and Stuart Wells. 1978. *The Costs of Educational Media: Guidelines for Planning and Evaluation*. Beverly Hills and London: Sage.

Lee, Kiong Hock. 1984. Universal Primary Education: An African Dilemma. Washington, D.C.: World Bank, Education Department.

————. Forthcoming. Further Evidence on Economies of Scale in Higher Education. *Comparative Education Review*.

Levin, Henry M. 1983. *Cost-Effectiveness: A Primer*. Beverly Hills: Sage.

Maynard, J. 1971. *Some Microeconomics of Higher Education: Economies of Scale*. Lincoln, Neb.: University of Nebraska Press.

McMeekin, R., and J. P. Gittinger. 1984. Cost-Effectiveness Comparison of Technical and Vocational Projects. World Bank Economic Development Institute Exercise Series no. 135/018. Washington, D.C.

Perraton, Hilary, ed. 1982. *Alternative Routes to Formal Education*. Baltimore, Md.: Johns Hopkins University Press.

Psacharopoulos, George. 1980. *Higher Education in Developing Countries: A Cost-Benefit Analysis*. World Bank Staff Working Paper no. 440. Washington, D.C.

————. 1982. The Economics of Higher Education in Developing Countries. *Comparative Education Review* 26, no. 2 (June):139–59.

Psacharopoulos, G., and W. Loxley. Forthcoming. *Diversified Secondary Education and Development: Evidence from Colombia and Tanzania*. Baltimore, Md.: Johns Hopkins University Press.

Riew, O. 1966. Economies of Scale in High School Operation. *Review of Economics and Statistics* 48, no. 3 (August):280–87.

Sirken, Irving, ed. 1979. Education Programs and Projects: Analytical Techniques, Case Studies and Exercises. Washington, D.C.: World Bank, Economic Development Institute.

————, ed. 1982. Education Programs and Projects: Analytical Techniques, Solutions. Washington, D.C.: World Bank, Economic Development Institute. Processed.

Spain, Peter L., Dean T. Jamison, and Emile G. McAnany, eds. 1977. *Radio for Education and Development: Case Studies Vol. 2*. World Bank Staff Working Paper no. 266. Washington, D.C.

Speagle, R. E. 1972. *Educational Reform and Instructional Television in El Salvador: Costs, Benefits and Pay Offs*. Washington, D.C.: Academy for Educational Development.

Tibi, C. 1983. Les déterminants des coûts de l'éducation. Paris: International Institute for Educational Planning. Processed.

Unesco. 1977. *The Economics of New Educational Media. Vol 1. Present Status of Research and Trends*. Paris.

———. 1980. *The Economics of New Educational Media. Vol. 2. Cost and Effectiveness*. Paris.

Wagner, Leslie. 1982. *The Economics of Educational Media*. London: Macmillan.

World Bank. 1978. Report of the External Advisory Panel on Education to the World Bank. Washington, D.C.: Education Department.

Zymelman, Manuel. 1976. *Patterns of Educational Expenditures*. World Bank Staff Working Paper no. 246. Washington, D.C.

———. 1982a. Educational Expenditures in the 1970s. Washington, D.C.: World Bank, Education Department.

———. 1982b. A Simulation Model for Forecasting Teachers' Salaries. Washington, D.C.: World Bank, Education Department.

8

Internal Efficiency and Educational Quality

Efficiency is a term used to describe the relationship between inputs and output, but because this relationship can be analyzed from several perspectives, judgments about efficiency may have to take into account more than one aspect of the relationship. Investment decisions, for example, need to consider both *external* and *internal* efficiency. The problem is that educational output is too complex to allow us to adopt a single index of either external or internal efficiency.

As we have already seen, the objectives of society are used to measure external efficiency, which can be judged by the balance between social costs and social benefits, or the extent to which education satisfies manpower and employment needs. More specifically, the external efficiency of schools may be judged by how well schools prepare pupils and students for their roles in society, as indicated by the employment prospects and earnings of students. Such measures depend on external criteria rather than on results entirely within the school.

In contrast, internal efficiency is concerned with the relationship between inputs and outputs within the education system or within individual institutions. Output in this case is measured in relation to internal institutional goals rather than the wider objectives of society. Clearly, the two concepts are closely linked, but it would be possible to envisage a school that was extremely efficient in developing skills and attitudes that were not highly valued in society as a whole. In such circumstances, the criteria of internal and external efficiency would conflict, and the school would be judged to be internally efficient but externally inefficient.

Since internal efficiency is measured in relation to the objectives of education, judgments about efficiency will depend on the way educational output is defined and measured. In other words, the quality as well as the quantity of inputs and outputs will have to be considered. The quality of output, however, is hard to measure where education is concerned. In this chapter we examine various methods of measuring both the quantity and quality of output.

To assess the internal efficiency of education, we need a statement of its aims and objectives together with a range of measures of output that reflect these various objectives and the success with which they are achieved. It is, of course, extremely difficult to measure the success with which the wider objectives of education are achieved, but some analysts have used such measures as examination scores; cognitive tests in a wide range of subjects; the length of time needed for pupils to reach a required standard; scores on standardized tests of reading ability and of language, mathematics, and science skills; and noncognitive tests designed to measure pupils' attitudes and motivation (for example, their attitudes toward modernity).

Another distinction that has great bearing on investment decisions is that between *technical efficiency*, which is concerned with the maximum output that can be achieved from a particular input of resources with a given level of technology, and *economic efficiency*, which is concerned with achieving a desired level of output at minimum cost. Studies of efficiency in the field of education are concerned with cost-effectiveness and economic rather than technical efficiency, although some studies of the effects of new technologies such as television or computer-assisted learning are primarily concerned with technical efficiency.

Economic or technical efficiency is analyzed by examining the way inputs are transformed into outputs—which in economic parlance is called the production function. The production function of education is imperfectly understood, however, since education is extremely complex and a great many variables affect the quality of output, including socioeconomic and family influences as well as school inputs.

Educational output is difficult to measure and is affected by so many factors that some have questioned the value of measuring and analyzing the internal efficiency of education. Some argue (Vaizey et al. 1972, for example), that efficiency and productivity are inappropriate concepts to apply to education as a whole. Others argue that nonschool factors such as family background or motivation can influence educational achievement to such a great extent that they may swamp the effect of school inputs.

Despite this pessimism, progress has been made in recent years in the measurement and analysis of internal efficiency, the relationship between school inputs and outputs, and the cost-effectiveness of education. There has been progress too in measuring some aspects of school output and quality, as well as in identifying factors that have considerable impact on achievement. We start, therefore, with the question of how to measure educational output.

Quantitative Indicators of Educational Output

The first point to clarify is what is meant by output. The World Bank (1980, p. 32) distinguishes between *output* in the sense of achievement of pupils or students—which refers to knowledge, skills, behavior, and attitudes—as measured by tests, examination results, and the like, and *outcome* in the sense of the external effects of output—that is, the ability of people to be socially and economically productive. We have already discussed the problems of measuring outcomes in terms of the extra earnings of educated workers or in terms of their employment and labor market performance. Such measures are important in assessing the external efficiency of education, but they are admittedly crude proxy measures for the long-term economic and social effects of education.

Similarly, pupils' scores in cognitive and noncognitive tests are used as proxies for the knowledge and skills learned in school—which constitute the output or value added of the education process—but because such tests may be difficult or expensive to administer or may be regarded as unreliable measures of quality, analyses of internal efficiency often measure output in purely quantitative terms, such as the number of graduates or qualified school leavers produced in the education system.

For many purposes, measuring output by the number of students who successfully complete a course provides a good first approximation, and comparing this with input measured by the number of pupil-years may be enough to indicate that high rates of wastage and repetition are connected with very low efficiency, judged simply by the length of time needed to produce one qualified school leaver. The World Bank (1980) estimated input-output ratios for fifty-four developing countries in 1970–75 (input is measured by actual pupil-years and output by school completers multiplied by the normal length of schooling); a ratio of 1 indicates maximum efficiency, but an index of 5, as in Burundi, indicates considerable inefficiency and wastage (table 8-1). It means, in fact, that five school places have to be provided for every successful primary-school leaver because of repetition and dropout. In many developing countries, less than half of the pupils enrolled in the first grade of primary school reach the fifth grade; the remainder will drop out without attaining permanent literacy, and many of those who do complete primary schooling will repeat one or more grades.

Because this kind of wastage is widespread in developing countries, the term *internal inefficiency* is sometimes taken to mean simply the rate of

Table 8-1. *Student Input-Output Ratios of Fifty-four Developing Countries, 1970–75*

Income group	Number of countries	Median	High	Low
Low	17	1.98	5.16 (Burundi)	1.20 (Kenya)
Lower middle	13	1.67	2.03 (Thailand)	1.14 (Jordan)
Intermediate middle	16	1.48	2.50 (Dominican Republic)	1.03 (Korea, Rep. of)
Upper middle	8	1.30	2.38 (Gabon)	1.12 (Singapore)
Total	54	1.65	5.16	1.03

Note: Input is measured by actual student-years. Output is measured by school completers multiplied by the normal length of schooling. A ratio of 1 indicates maximum efficiency; a ratio of greater than 1 indicates wastage.
Source: World Bank (1980), p. 30.

retention, and the ratio between the actual number of years needed to produce a qualified school leaver and the normal school cycle (the input-output ratio shown in table 8-1) is sometimes referred to as an inefficiency index. Certainly one way to improve the internal efficiency of education in developing countries would be to reduce wastage and repetition, although we argue later in the chapter that quality as well as quantity must be improved.

Wastage and Repetition

From the extensive research on wastage and repetition that has been carried out by Unesco (1977, 1982) and the International Bureau of Education (1971, 1972), and that is summarized in Haddad (1979), it is evident that the problem of dropout and repetition of grades is serious throughout the developing world. Estimates of repetition rates in selected countries in the early 1970s indicate that a third or even a half of all pupils in many developing countries repeat the first grade, and a quarter or more pupils repeat subsequent grades (table 8-2). According to a Unesco (1977) survey of repetition in developing countries, repeaters constitute about 15 percent of total enrollment in primary education in Latin America, 15 percent in Africa, and 18 percent in South Asia. In other words, if there had been no repetition, the number of children of primary-school age admitted to school could have been increased by some 15–20 percent without extra expenditure. Furthermore, it has been estimated that the total amount devoted to repeaters in the first grade in Latin America is more than US$300 million a year (Schiefelbein 1975).

Dropout and repetition appear to be most common among students from a low socioeconomic background and are more prevalent in rural than in urban areas, and among females than among males. Causes include poverty, which may give rise to illness, malnutrition, and absenteeism; the high opportunity cost of schooling for poor families (discussed in chapter 5); cultural factors, which affect girls in particular; inappropriate curriculum and examinations, which are often excessively academic and designed to prepare a minority of pupils for upper secondary and higher education; badly trained teachers; lack of textbooks and materials; overcrowded schools; and a shortage of secondary-school places, which leads to repetition at the primary level. Dropout and repetition have been identified as a principal cause of internal inefficiency in Morocco, where only 24 percent of pupils complete the five-year primary cycle without repeating and 21 percent drop out before completing the cycle; in Mauritius, where more than 25 percent of all primary-school pupils repeat grades; and in the Dominican Republic, where as many as 60 percent of primary-school pupils in 1970 had to repeat a grade. Such high rates of wastage mean that the average number of years required to produce one primary-school completer is not five or six, as in the normal primary cycle, but eight or ten years. Since every repeater displaces a potential new pupil, many countries that have not yet achieved universal primary education (UPE) could increase enrollment ratios to 100 percent by eliminating repetition in primary school classes and would not have to increase facilities or teaching staff.

The following are a number of steps that could be taken to improve the flow of pupils through primary and secondary schools in many developing countries:

- Introduce automatic or semi-automatic promotion between grades. Several studies reviewed by Haddad (1979) have suggested that there is no educational advantage to be derived from making low achievers repeat grades.
- Change examination procedures or reduce the number of examinations taken. Either would reduce repetition and dropout.
- Improve the quality of the curriculum and make it more relevant to pupils' interests and surroundings.
- Improve basic teaching equipment. A World Bank project in Morocco, for example, is expected to help reduce wastage by providing teaching aids for 700 primary schools.
- Improve teacher training so that empirical learning will replace memorization and rote learning.
- Strengthen pedagogical research as an instrument for improving educational efficiency.
- Provide more training opportunities for primary-school leavers in order to reduce repetition of the final grade.

Table 8-2. *Rates of Repetition in Selected Countries, 1970–76*
(percent)

Region/country	Year	\multicolumn{6}{c}{Grade in first cycle}					
		1	2	3	4	5	6
Asia							
Burma	1973	25	20	19	18	15	...
India	1970	26	20	18	17	16	...
Indonesia	1975	16	11	11	9	7	2
Korea, Rep. of	1975	...	...	...	...	...	...
Malaysia	1974	...	...	...	...	...	...
Papua New Guinea	1972	74	3	2	2	1	2
Singapore	1974	...	...	3	4	4	27
Sri Lanka	1976	14	12	12	12	18	...
Thailand	1975	19	12	11	4	7	4
East Africa							
Botswana	1975	...	...	...	...	...	...
Burundi	1974	19	18	25	26	31	39
Kenya	1975	5	3	4	5	4	7
Lesotho	1975	9	4	4	3	3	3
Madagascar	1973	24	15	14	15	10	16
Malawi	1974	19	17	15	12	10	13
Rwanda	1974	27	18	17	21	18	24
Somalia	1974	...	...	...	...	...	...
Sudan	1971	...	...	4	4	3	9
Swaziland	1975	9	6	8	11	11	11
Tanzania	1971	1	1	1	3	...	...
Zaire	1971	27	24	23	19	19	18
Zambia	1974	...	1	1	4	1	1
West Africa							
Benin	1974	17	16	20	19	27	36
Burkina	1974	14	12	14	14	19	36
Cameroon	1973	32	24	25	20	23	35
Central African Republic	1974	32	37	24	26	29	36
Chad	1975	41	34	32	28	32	57
Congo	1974	32	26	31	26	25	34
Gabon	1974	50	33	29	21	21	23
Gambia, The	1975	7	4	4	5	6	35
Ghana	1975	5	2	2	2	1	1
Ivory Coast	1973	21	21	22	21	27	46
Mali	1970	25	23	27	23	29	37
Niger	1974	8	13	17	15	19	36
Senegal	1972	11	14	15	16	18	34
Togo	1975	37	24	26	20	25	34

Table 8-2 *(continued)*

		Grade in first cycle					
Region/country	*Year*	*1*	*2*	*3*	*4*	*5*	*6*
Europe, Middle East,							
and North Africa							
Algeria	1975	7	9	13	12	13	21
Egypt	1974	...	14	...	15	...	13
Greece	1973	8	5	4	2	2	1
Iran	1970	11	11	8	7	7	9
Iraq	1973	20	16	13	17	32	18
Jordan	1974	...	...	7	8	7	4
Kuwait	1974	14	15	16	12	...	...
Morocco	1975	23	20	26	30	44	...
Oman	1974	6	9	12	8	7	4
Saudi Arabia	1974	17	13	14	22	15	9
Syria	1974	12	12	10	9	7	8
Tunisia	1975	11	13	14	14	19	41
Yugoslavia	1973	7	5	4	3	7	7
Latin America and							
the Caribbean							
Argentina	1975	21	12	10	6	5	3
Brazil	1974	24	17	11	10	12	11
Chile	1974	19	15	13	10	9	8
Colombia	1974	20	16	13	11	8	...
Costa Rica	1975	...	13	14	...	8	3
Dominican Republic	1970	33	21	18	13	10	6
Ecuador	1973	17	14	11	11	10	9
El Salvador	1974	14	9	6	5	3	2
Guatemala	1972	26	14	12	7	5	2
Guyana	1973	18	7	7	6	8	14
Mexico	1965	22	14	13	11	9	4
Nicaragua	1973	16	11	9	8	6	3
Panama	1974	21	18	15	9	7	3
Paraguay	1974	25	20	16	11	7	3
Peru	1975	15	12	9	8	7	4
Uruguay	1973	30	18	16	14	13	7
Venezuela	1974	...	...	...	10	9	...

... Zero or negligible.

Note: Countries are grouped by region according to World Bank grouping of borrowing countries. "Asia" region includes both "East Asia and Pacific" as well as "South Asia."

Source: World Bank (1980), pp. 116–17.

- Introduce school meals. This step may reduce dropout due to malnutrition and ill health.
- Introduce or modify age restrictions governing admission to lower grades or progression through grades. Change financing policy that may be affecting dropout.

The World Bank has provided loan assistance for a number of projects designed to reduce wastage and dropout by improving the curriculum. These projects have included agricultural schools in Chile; diversified secondary schools in Colombia, Jordan, Nigeria, and Tanzania; industrial schools in Brazil; and agricultural schools in Indonesia. One way to evaluate the success of such projects is to compare wastage and repetition rates in project schools with those in general secondary schools.

Unfortunately, many developing countries have no accurate information on repetition or dropout rates. A device that can be used to estimate wastage is a flow chart that shows the progression rate between grades. This has been done in Colombia, for example, for comprehensive secondary schools (financed by means of World Bank loans in 1968 and 1970) and for other secondary schools (table 8-3). The purpose of the new secondary schools, which are known as INEMs, was to provide diversified secondary education that was more practically oriented and that could respond better to the needs of the labor market and the interests of individual students. Another goal was to reduce wastage. Progression rates in some INEM schools were considerably better than the average for all secondary schools in Colombia, but in many schools the enrollment in grade 11 was only 30 or 40 percent of the enrollment in grade 6 five years earlier (see table 8-3). This finding raises the question of whether the investment of additional resources to reduce wastage and repetition is likely to be profitable. A common justification for such investment is that it will significantly improve internal efficiency by increasing the proportion of pupils who complete schooling in the minimum required time. The profitability of the investment depends, however, on whether the extra cost per *pupil* would lead to a substantial saving in cost per *completer*. Berstecher (1970) shows that in some circumstances the gains in quantitative efficiency would be enough to justify extra expenditure per pupil. Like the effort to introduce curriculum reform in Colombia, most attempts to improve the internal efficiency of schools are directed toward qualitative as well as toward quantitative improvements.

Qualitative Indicators of Educational Output

To measure the achievement of pupils or students, we need a technique that captures both the quantity and quality of learning. The quality of

Table 8-3. *Standard Progression of INEMs
and the National Secondary System, Colombia*

Schools	Grade						Average
	6	7	8	9	10	11	
Kennedy (Bogotá)	1,000	976	885	771	715	649	833
Cali	1,000	791	691	749	656	601	748
Medellín	1,000	816	601	460	419	323	603
Tunal (Bogotá)	1,000	811	727	599	500	450	681
Barranquilla	1,000	740	525	411	291	274	540
Cartagena	1,000	828	608	472	424	355	615
Pasto	1,000	874	794	664	604	475	735
Bucaramanga	1,000	843	729	599	534	487	699
Santa Marta	1,000	770	627	480	422	343	607
Cucuta	1,000	671	515	445	392	379	567
Manizales	1,000	796	646	595	552	391	663
Monteria	1,000	644	556	411	338	269	536
Armenia	1,000	681	554	471	362	324	565
Tunja	1,000	812	664	552	472	430	655
Pereira	1,000	635	497	439	377	313	544
Popayan	1,000	770	658	427	377	315	591
Ibague	1,000	747	593	416	352	316	571
Neiva	1,000	883	658	583	478	428	672
Villavicencio	1,000	706	557	467	400	302	572
Average for 19 INEM	1,000	779	636	527	456	391	632
National system							
Public schools	1,000	672	526	460	385	382	570
All schools	1,000	691	542	501	398	396	588

Source: World Bank estimates.

output can be measured in two ways. One is to use examination scores as tests of attainment, as has been done in Kenya (Thias and Carnoy 1969), where scores in the Kenya Preliminary Examination (KPE) taken at the end of primary schooling were used as measures of pupil achievement. The alternative is to use specially administered tests that measure noncognitive as well as cognitive achievement. The disadvantage of this technique is that the tests are costly and therefore only small samples are used, with the result that reliability may be questionable.

One ambitious attempt to measure educational output by means of test scores has been the International Project for the Evaluation of Educational Achievement (IEA), which began by testing levels of achievement in mathematics in twelve countries (Husen 1967), followed by attainment in science in nineteen countries (Comber and Keeves 1973), and in reading comprehension in fifteen countries (Thorndike 1973); more recently, both the number of countries and the subjects surveyed have been increased (Peaker 1975). The tests were carefully constructed so that they conformed closely to curriculum objectives. Each test item was translated into the national language, and extensive pretesting was carried out to ensure international comparability of the tests. Other international studies have also used specially designed tests to compare pupil achievement in different countries. Argentina, Brazil, Bolivia, Colombia, Mexico, Paraguay, and Peru, for example, cooperated in a large-scale program that collected data on school variables such as expenditure, repetition, and dropout rates; at the same time, pupil achievement tests were conducted under the auspices of the Programa de Estudios Conjuntos sobre Integración Económica Latinoamericana (ECIEL).

The results of such international comparisons of achievement have proved fruitful for research on the influence of different factors on achievement, but inevitably they provide only a partial measure of the quality of output. The problem is that variations in quality of output cannot be measured as reliably or easily as variations in input. Many studies use class size or expenditure per pupil to measure quality, for example, but these factors cannot explain input-output relationships since the very question that needs to be answered is whether increasing inputs does lead to an increase in either the quantity or quality of output. Research on the effects of class size on pupil achievement (Haddad 1978) shows that there is no automatic link between class size and quality of

Table 8-4. *Relationship between Performance and Expenditure per Pupil, Malaysia*

Schools by category	Index of average performance		Average expenditure per pupil		Index of educational performance	
	Malay	English	Malay	English	Malay medium	English medium
Performance category						
Highest performance quartile	11.69	11.10	246	320	—	—
Mean, all schools	8.31	8.00	247	301	—	—
Per-pupil expenditure category						
Highest-cost quartile	—	—	—	—	7.62	7.87
Middle two quartiles	—	—	—	—	8.88	8.36
Lowest-cost quartile	—	—	—	—	7.91	7.47

— Not applicable.
Source: McMeekin (1975).

output. Furthermore, many input-output studies have found little connection between expenditure per pupil and achievement. In Malaysia, for example, secondary schools with the highest achievement scores do not spend more than average, and schools with the highest costs do not achieve above-average results (table 8-4). Clearly, inputs cannot be used as a measure of output quality in this case.

Among the most important inputs to the education process are the pupils or students themselves. To test whether variations in output, measured by examination results or test scores, simply reflect variations in the quality of student input, a few studies have tried to identify value added by comparing test scores at the start and completion of a course. In Tunisia, for example, the grade point average (GPA) in the secondary-school entrance examination of a sample of secondary-school pupils was compared with their GPA at the time of the survey (Carnoy and Thias 1974). Few such attempts have been made in other developing countries, however.

Input-Output Relationships in Education

The relationship between inputs and outputs of education—which, as noted earlier, is sometimes called the education production function—is highly complex since many factors (such as ability, home background, and socioeconomic factors) besides school variables affect educational outcomes. The term production function refers to the process by which inputs are converted to outputs. A simple production function for education would be $A = f(T, B, E . . .$ and so on) where A = achievement, T = teacher-pupil ratio, B = books and other materials, E = equipment, and so on. Experience shows, however, that the education production function is far more complex than this and includes many more variables. An early study (Alexander and Simmons 1975) of the educational production function in developing countries examined the relationships and interactions between preschool factors, schooling, educational outputs, and final outcomes (see figure 8-1). On the basis of studies in the United States, it suggested that family background and socioeconomic factors are more important determinants of pupil achievement than school variables such as teacher qualifications or expenditure on books.

In the early 1970s much of the research on the inputs and outputs of schooling in the United States concluded that school variables had little effect on educational outcome (Averch et al. 1974). The much publicized findings of the Coleman Report (Coleman et al. 1966), that socioeconomic factors are more important than school variables in explaining regional and racial differences in pupil achievement, were followed by

Figure 8-1. *The Learning System: Causes, Consequences, and Interaction*

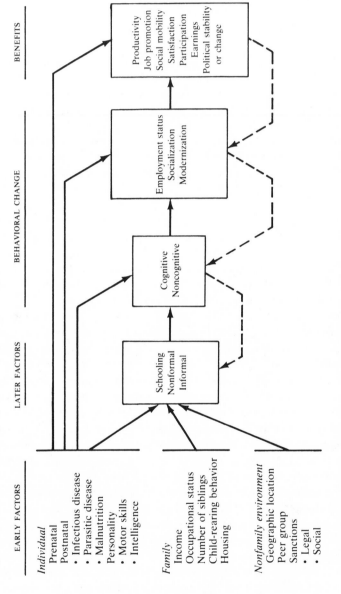

Note: Other arrows are omitted to maintain the clarity of the diagram. For example, Family and Nonfamily Environment should have dotted lines to Later Factors and Behavioral Change.
Source: Alexander and Simmons (1975).

the even more negative conclusion of Jencks (1972, p. 256): "The characteristics of a school's output depend largely on a single input, namely the characteristics of the entering children. Everything else, the school budget, its policies, the characteristics of the teachers—is either secondary or completely irrelevant."

This pessimism is reflected in the titles of some of the research reviews that appeared at this time, for example, "Do Teachers Make a Difference?" (HEW 1970), and "How Effective is Schooling?" (Averch et al. 1974). The answers, according to many of the studies, were "not much" and "not very." This pessimism also pervades Alexander and Simmons's (1975) review of the relationship between input and output in nine developing countries (Chile, Congo, India, Iran, Kenya, Malaysia, Puerto Rico, Thailand, and Tunisia). They concluded that schooling inputs have only a weak or insignificant impact on pupil achievement, whereas other determinants such as home background and individual personality have a stronger influence. The only school variables they found to be related to internal efficiency of education in developing countries were teacher motivation (rather than experience or qualifications), textbooks, and other reading materials. This finding has disturbing implications for any attempts to increase the efficiency of education by changing the combination of inputs or improving teacher quality or school facilities.

This pessimistic conclusion has been challenged, however, by further research. Studies in more than twenty developing countries, including India (Shuluka 1974) and Uganda, Kenya, Ghana, and Papua New Guinea, as well as evaluations of World Bank education projects in Kenya and Somalia (Heyneman 1980; Heyneman and Loxley 1983) all conclude that wealthy school children do not perform better in achievement tests, and thus suggest that socioeconomic background has much less effect on pupil achievement in developing than in the developed countries. In fact, a more recent review of the factors affecting the academic achievement of school children (Schiefelbein and Simmons 1978) concludes that the less developed a society, the smaller the influence of home background on achievement and the greater the effect of school variables. It has been shown in a sample of twenty-nine countries, for example, that the proportion of explained test score variance attributable to school quality is lowest in developed countries such as Australia, Japan, Sweden, and the United States, but it is twice or three times as high in Brazil, Botswana, India, or Thailand (figure 8-2).

As a result of such findings, pessimism about the influence of school variables and the scope for improving internal efficiency by changing the combination or quality of inputs has been much reduced. It is estimated that the gains from improving internal efficiency in Malyasia, for exam-

Figure 8-2. *Relationship between Per Capita Income and the Effects of Primary-School Quality on Academic Achievement*

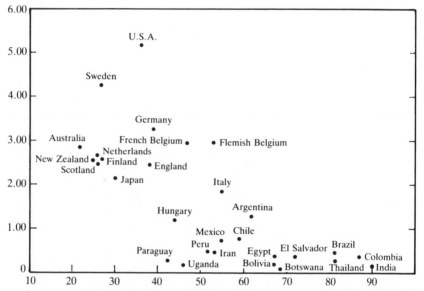

Per capita GNP (thousands of 1971 U.S. dollars)

Percentage of explained academic achievement variance attributable to school quality

Note: $r = -0.72$, $P < 0.001$, $N = 29$. Academic performance was measured by mathematics tests in Botswana, Uganda, and Egypt and by science tests in all other countries.
Source: Heyneman and Loxley (1983)

ple, are as great as the potential gains from external efficiency (McMeekin 1975; Beebout 1972). World Bank research on the determinants of academic achievement demonstrated that variations in inputs do affect educational outputs. Among the most important factors are teachers and textbooks.

Although research on class size has not demonstrated a clear relationship between teacher-pupil ratios and achievement (Haddad 1978), a review of thirty-two studies in developing countries found evidence that teacher qualifications are important, and concluded that "contrary to the arguments presented elsewhere, the evidence here suggests that trained teachers do make a difference" (Husen, Saha, and Noonan 1978, p. 42). In addition, Heyneman, Farrell, and Sepulveda-Stuardo (1981) concluded from studies in ten countries that achievement is more closely correlated with textbook availability than with other measures of school inputs, such as class size or expenditure on teacher salaries.

Other school variables have also been shown to have an impact on pupil achievement. In Peru, for example, school management variables are important along with teacher attributes and physical factors, which would include availability of libraries, visual aids, and basic equipment such as tables and chairs (Arriagada 1983).

As a result of such findings, the focus of World Bank research has shifted from the question of *whether* investment in school inputs can increase outputs and internal efficiency, to *what* is the most cost-effective way of improving inputs. Issues of concern now, for example, are whether it is better to improve teacher training or increase textbooks and teaching materials in order to improve the output of schools. Some studies have also been concerned with the way pupils acquire and retain skills.

A study of literacy and numeracy among Egyptian primary school pupils (Hartley and Swanson 1984) used the concept of learning and retention curves (which measure skill acquisition and retention) to explore the question of whether distinct thresholds mark the achievement of permanent literacy, and whether dropouts incur a significant loss or wastage of skills after leaving school. During this study, specially designed tests were given to a sample of nearly 5,000 pupils or former pupils who had dropped out of Egyptian primary schools. The scores on the achievement tests were compared to show how skill levels vary among school pupils and dropouts over time with respect to sex, urban or rural community, and other such categories. The changes in skill levels over time represent learning and retention curves for each skill. The results show that retention curves of dropouts are very low and consistently below the level of functional literacy or numeracy. This suggests that there is no single "retention threshold" among dropouts and that if they are to attain functional literacy or numeracy, special measures such as adult literacy instruction may be necessary.

The Effect of Teacher Training

The belief that investment in teacher training will improve the quality of schooling by increasing the level of pupil achievement led the World Bank to emphasize teacher training facilities in lending for education projects during the 1960s and early 1970s. This belief was questioned, however, when research cast doubt on the importance of teacher training in determining pupil achievement. A study in Latin America, for example, suggested that students do almost as well when studying under teachers trained in a normal school as they do when taught by university graduates (World Bank 1974).

By the mid-1970s many in developed countries had become skeptical about the importance of teacher training: "Research to date has found little or nothing in school resources that consistently and unambiguously makes a difference to student outcomes, such as achievement. Once widely held beliefs about the overriding importance of school and teacher quality have been called into question, if not refuted" (Nollen 1975, p. 74). To examine this issue further, the Bank commissioned a review of research on the relationship between the quality and training of teachers and student achievement. The skepticism, it found, was not justified by the experience of developing countries. After examining thirty-two studies, the review concluded that trained teachers do make a difference, and in particular that teacher qualifications, experience, and amount of education and knowledge are positively related to student achievement (Husen, Saha, and Noonan 1978). This conclusion was reinforced by further analysis of data on student achievement in science that had been collected by IEA in nineteen countries during 1970 (Comber and Keeves 1973). The analysis suggested that if the level of teacher training in Chile and India was increased, the average test scores of pupils in secondary schools would improve, and it concluded that investments in teacher training programs would help to improve the quality of output in terms of student cognitive scores.

The question remains, of course, whether improvements in teacher training are more cost-effective than other improvements in inputs, and what form they should take—an upgrading of existing teachers, in-service training for poorly trained teachers, retraining of teachers for new curricula, or improvements in initial training. Unfortunately, there is little evidence to indicate which of these alternatives would be the most efficient. The review therefore concludes with a plea for more research on cost-effectiveness, with a change of emphasis: "Future research should not be preoccupied with the question of *whether* trained teachers make a difference, since that question has already been answered by cumulative research evidence. The question which remains unanswered is *how*, and *because of what qualities and in what contexts* do teachers make a difference" (Husen, Saha, and Noonan 1978, p. 47).

Research is continuing in this area. Some studies suggest that teacher experience is important in primary and lower secondary education, but that the skill and knowledge of teachers, as reflected in their salary or qualifications, are more important in the higher grades. It is not clear, however, whether there are upper or lower thresholds for teacher qualifications. The evidence suggests that such thresholds exist but that they are likely to vary among developing countries.

Research on the links between teacher training and pupil performance does not, therefore, provide a model for developing countries to copy. It shows that teachers do make a difference and that the quality of teachers is important, but it also shows that the best way to improve teacher quality will depend on conditions in the country and can be determined only after analysis of the costs and effectiveness of alternative ways of training and using teachers. It may also depend on the provision of textbooks and other complementary inputs.

The Influence of Textbooks

Whereas evidence on the effects of class size and teacher variables on student achievement is often conflicting or ambiguous, the evidence of a relationship between the provision of books and achievement is clear and consistent. A review of studies in ten developing countries reported a more consistent relationship between pupil achievement and the availability of books than between achievement and other variables such as teacher training, class size, teacher salaries, boarding facilities, grade repetition, and so on (Heyneman, Farrell, and Sepulveda-Stuardo 1978).

Although this review points to the likely returns from investment in textbooks in developing countries, many unanswered questions remain. How, for example, do textbooks interact with other variables, particularly teacher quality? Are textbooks more important in large classes? How cost-effective are different types of textbooks? (How do cheap mimeographed materials compare with more expensive, permanent books, for example, or how do locally produced materials compare with standard textbooks?) Are books more important for some subjects than others? Are they more important for inexperienced teachers? Are they more important for certain pupils? Research in Chile and Mexico and World Bank experience in the Philippines has shed some light on these questions.

Some surprising results were obtained in Chile, where Schiefelbein, Farrell, and Sepulveda-Stuardo (1983) looked specifically at the attitudes of teachers and their use of textbooks in both public and private schools, the influence of teacher experience and training, and the extent of subject differences in the use of textbooks. Far from showing, as had been suggested, that young, inexperienced teachers are more likely to use textbooks to overcome their lack of experience, the survey showed that the reverse was true: less experienced teachers are less likely to use textbooks than those with more experience, and 78 percent of all the teachers in the survey expressed negative or ambivalent attitudes toward the use of textbooks. There were also significant differences between subject areas. Science and mathematics teachers were less likely to use textbooks, and the proportion of teachers who "never" used textbooks was much higher in science (46 percent) and mathematics (33 percent) than in English (4 percent) or a language. One reason for the lack of enthusiasm for textbooks may be a lack of emphasis on textbooks in teacher training. More than half the teachers said they had not had any training in the use of textbooks. The study therefore concluded that perhaps one of the most important steps that should be taken before free textbooks are provided in public schools is simply to prepare teachers to use them.

A survey among school pupils suggested that students have a much more positive attitude toward textbooks than do teachers, but when teachers recommended that pupils should buy their own books, a significant proportion reported that they could not afford to do so. Thus, the main problems in Chile with respect to textbook availability and utilization have to do with teacher attitudes and the inability of poorer children to buy books, even when teachers wish to use them. More than half of all students surveyed did not possess a textbook, even though the government of Chile subsidizes book production; a report in 1971 noted that even in very remote and poor villages, local kiosks had textbooks for sale (Heyneman, Farrell, and Sepulveda-Stuardo 1978).

The question to ask at this point is whether governments should provide free textbooks. In Mexico, the provision of free textbooks for primary-school pupils is an important part of the government's policy to improve educational efficiency and equity; the National Commission for Free Textbooks (CNLTG) was set up in 1959, and by 1981 every primary-school child in Mexico had free textbooks. The Mexican government believes that the use of textbooks raises academic standards and increases the efficiency of a school system, and that the production and distribution of textbooks are as important as teachers' salaries and school buildings in determining output. Although in 1980 government subsidies on books amounted to only 1 percent of the total education budget, the government believes this is one of the most essential items in it.

The Mexican experience has been summarized by Neumann and Cunningham (1982), who concluded that textbook development and supply require and deserve the same priority as teacher development and school construction; they argue that textbooks require a long-range government commitment supported by regular and adequate annual expenditures. Adequate funding can be provided by setting aside a small percentage, say, 2–4 percent, of the national budget for education. Whether textbooks should be published by government or by private sector publishers is a policy choice. In Mexico, the government nationalized primary-school textbook publishing in 1959, but now recognizes the importance of private publishing and provides financial assistance for the publishing industry. Neumann and Cunningham also emphasize that printing is not publishing, and the two should not be confused; publishing need not be carried out where the books are to be printed. Finally, the study concludes that there are strong arguments for providing every child in the school system with free books, but choice is important both for teachers and for children of differing backgrounds and abilities, so that every effort must be made to build choice into a national free textbook program.

Although the nationalization of textbook publishing is believed to have important benefits in Mexico, it is not necessarily appropriate for all developing countries. According to Heyneman, Farrell, and Sepulveda-Stuardo (1981), centralization does not necessarily lead to greater efficiency or equity. Uganda centralized the purchasing of school textbooks in 1972 in order to improve efficiency of production and the distribution of books, but the relationship between pupils' socioeconomic status and access to books was actually strengthened rather than diminished by the centralization (Heyneman, Farrell, and Sepulveda-Stuardo 1981, p. 245).

In another interesting case, the Philippine government, with the assistance of the World Bank, launched a US$37 million textbook project in 1977–78 to provide improved textbooks and to increase the ratio between textbook and pupils from 1:10 to 1:2 or, in a subsample of schools, 1:1. Teacher training in the use of textbooks was also an integral part of the project. The effects were monitored and pupils given achievement tests before and after the project, which initially concentrated on science, mathematics, and language textbooks. According to the cognitive test scores, the increase in the number of textbooks had a sizable impact on pupil achievement in these three areas. Moreover, the gains were not confined to a few experimental classrooms, but were observed all over the country (Heyneman, Jamison, and Montenegro 1984).

Further improvements were not observed, however, when a 1:1 textbook-pupil ratio was achieved instead of the 1:2 ratio. In any case, to provide one textbook for every child, instead of one to be shared between two pupils, would have doubled the cost of the project. Therefore, on the grounds of very simple cost-effectiveness analysis, the additional investment would not be justified, whereas the provision of books on a one-to-two basis seems to be a highly worthwhile investment, since the textbooks require only a 1 percent increase in the annual expenditure on education per student.

The results in the Philippines strongly support the belief of the Mexican government that textbooks are often the most cost-effective means of improving academic achievement and increasing the efficiency of schools. The experience in the Philippines also suggests that learning gains are frequently greatest among the poorest or most disadvantaged pupils, and thus that there are no potential conflicts between efficiency and equity. Furthermore, the evidence from both countries, together with that from small-scale studies in other countries—such as a study of mathematics teaching in Nicaragua (Jamison et al. 1981)—suggests that investment in books may significantly improve the efficiency of education, particularly at the primary level.

At the same time, these projects show that it is not enough simply to provide textbooks. Some effort must also be made to ensure that they are adequately used. Both in the Philippines and in Indonesia, where a project financed with World Bank assistance involved the printing and distribution of 138 million textbooks, there is evidence that too much attention was paid to the production and distribution of the books, and not enough to discovering how relevant they were to classroom needs and how effectively they were used (Neumann 1980). In other words, not enough attention was paid to quality control and institution building as objectives of textbook and publishing projects.

At present, developing countries devote a very small proportion of school expenditure to teaching resources, including books, maps, or visual aids. Industrial countries allocate 14 percent of primary-school recurrent costs to classroom resources (books, teaching aids, furniture, and so on) and 86 percent to salaries, whereas the average in Asia is 9 and 91 percent, and in Africa 4 and 96 percent. Thus even a small reallocation of resources could increase efficiency; in fact, it has been suggested that a minimum of 10 percent of public recurrent expenditures should be devoted to teaching tools (Heyneman, Jamison, and Montenegro 1984). World Bank lending today reflects this change in emphasis. Whereas not one of the thirty-one education projects appraised by the Bank between 1963 and 1969 contained specific support for classroom materials, the provision of classroom materials has been a principal component in several projects since 1976.

If internal efficiency is to be improved through the reallocation of resources, the costs and effectiveness of alternative combinations of inputs must be assessed. As we have seen in the Philippines and in Nicaragua, examples of simple cost-effectiveness analysis can be found in developing countries, but systematic comparisons of the cost-effectiveness of alternatives are still rare.

Cost-Effectiveness Analysis

According to the evidence discussed so far, schools clearly *do* make a difference, and many types of inputs do determine educational outputs. The problem therefore is to identify the most cost-effective way to increase inputs or to change the combination of inputs so as to maximize output. Cost-effectiveness analysis, which proceeds by comparing the output achieved with various combinations of inputs, thus allows us to identify the lowest cost of achieving a desired level of output or effectiveness, or the greatest level of output or effectiveness that can be achieved for a given cost.

If the costs of two projects are the same, cost-effectiveness analysis can identify which achieves the most output, measured both quantitatively and qualitatively. If, however, effectiveness is the same, in terms of equal standards of achievement, the purpose of the analysis is to identify the least costly alternative. When alternatives differ in both their costs and effectiveness, cost-effectiveness analysis alone does not provide a basis for choice, although it may be helpful in showing the extra cost of achieving an increase in output.

Suppose, for example, that the cost per student in a general secondary school is $100 and the cost in a technical school is $150, but technical students achieve test scores 20 percent higher than those of general-school students. In this case, the two schools have different cost-effectiveness ratios, and cost-effectiveness analysis will not demonstrate which is the more cost-effective use of resources. This will depend on a judgment about whether a 20 percent improvement in achievement is worth $50 per student.

In practice, many examples of cost-effectiveness analysis in education fall into this category because it is difficult to ensure that the quality of output of two alternatives will be the same. Cost-effectiveness analysis may nevertheless be useful, not in providing conclusive proof of the superiority of one alternative, but in providing a framework for the systematic comparison of the costs of alternative ways of producing the same *quantity* of output, and for testing the sensitivity of the conclusions to different assumptions about *quality*.

As an example, cost-effectiveness analysis could be used to compare the costs of vocational education provided in a specialized vocational school and an alternative on-the-job training scheme that combines formal training with productive work (Zymelman 1973; McMeekin and Gittinger 1984). The costs of the two alternatives can be compared in relation to annual recurrent costs per student and annualized capital costs (see chapter 7). The quantitative output of the two training systems can be compared in relation to number of trainees. If the output of the two schemes is the same, then cost-effectiveness analysis consists simply of identifying which has the lowest costs. Differences in the quality of training may be significant, however, and Zymelman suggests measuring quality differences by means of employers' estimates of the quality of training and the probability that trainees will find employment after completing their training. A survey of employers could provide information about their subjective judgments on quality, but to obtain information about the employment prospects of graduates some form of tracer study would be needed to follow a sample of trainees and to collect data about their labor market experience, including earnings.

In the absence of data on the employment and earnings of workers with different types of training, subjective judgments about quality can be incorporated into cost-effectiveness analysis through a system of weights, and this technique may help to show whether the conclusions are sensitive to judgments and evaluations of quality. Thus, although the quality of output in alternative projects may be impossible to measure precisely, a simple type of cost-effectiveness analysis may still show the effect of alternative assumptions about quality. This technique has been used to compare different forms of vocational training in several developing countries. One of the most fruitful applications of cost-effectiveness comparisons has been in choosing between traditional educational methods and new media such as broadcasting. A review (Unesco 1982) of educational technology projects in eight developing countries concluded that technology can help reach more students, can provide equal cognitive opportunities, and, under certain circumstances, can increase learning and student motivation.

Another review of educational technology projects concluded that information on outputs or effectiveness is often fragmentary and inconclusive (Jamison and Orivel 1982). In Korea, for example, the fact that students of the Air Correspondence High School (ACHS) achieve less well on cognitive tests than conventional high school pupils may be due to their lower initial achievements. Thus in terms of educational "value added," the effectiveness may be greater, and in terms of equity (that is, opening opportunities for disadvantaged adults) the correspondence school performs well (Lee 1981). In the Dominican Republic, students who study through radio achieve the same test scores as conventional students, but the costs are much lower.

The overall cost-effectiveness of the projects reviewed by Jamison and Orivel generally compares well with that of traditional teaching. In many cases, however, cost-effectiveness depends on economies of scale. Lee, Futagami, and Braithwaite (1982), for example, explore whether it would be profitable for the Korean Educational Development Institute (KEDI) to invest in its own radio transmission network to increase the efficiency of ACHS. Such an investment would involve greater capital costs of broadcasting initially, but as broadcasting hours increased, the operational costs would fall. The relative costs of leasing time from existing networks and establishing KEDI's own network can be compared on the

Figure 8-3. *Comparison of the Present Value of Costs of Alternative Methods of Radio Education*

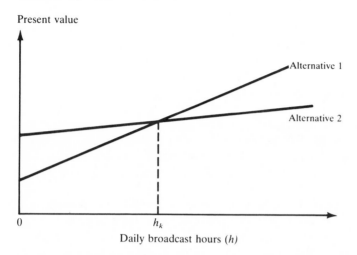

Present value

Alternative 1

Alternative 2

0

h_k

Daily broadcast hours (*h*)

Source: Perraton (1982), p. 166.

basis of the present value of costs (see figure 8-3). According to this type of analysis, there is a critical level of broadcasting hours (h_k) at which the second alternative becomes most cost-effective. If the broadcasting policies are equally effective with respect to student learning, the critical level would be four hours of broadcasts per day. Since programs on a special educational network could be broadcast at more convenient hours, however, the effectiveness of the second alternative could be increased and thus the critical point (h_k) reached earlier.

The size of the potential audience is therefore a critical variable in determining the relative cost-effectiveness of traditional teaching and the new media. One review of various education and communication technology projects in developing countries (McAnany 1980) emphasizes, however, that noneconomic factors have great bearing on the success of new technology projects. For example, the political commitment of a government to change and development is one of the most obvious, yet often neglected, factors explaining the success of certain educational technology projects, as has been demonstrated by radio campaigns in Tanzania in 1973 and 1975 and by educational television projects in El Salvador and Mexico. Another influential factor may be the degree of motivation of the target audience.

It is difficult to assess the precise contribution of such factors, but they have contributed to the success or failure of educational technology projects in many countries (including Malaysia, Thailand, and the Philippines). To maximize the potential of mass media for improving internal efficiency and effectiveness, it is essential to integrate them closely with curriculum improvement: "The idea is not to add on the educational media but to use them to upgrade the total educational experience" (Futagami 1981, p. 253).

Many cost-effectiveness studies have evaluated educational technology projects *after* they were introduced, but in the appraisal of investment projects such estimates are generally needed *before* a decision is made. Many developing countries want to improve teacher training, for example, and would like to be able to compare the relative efficiency of full-time initial training for teachers and in-service upgrading of existing teachers. Can cost-effectiveness analysis be used to throw light on this widespread problem?

Traditional patterns of teacher training are often conservative and expensive in relation to local needs and resources. Unfortunately, there is little information on the costs and effectiveness of alternative methods of teacher training, although two studies of in-service upgrading programs in Botswana (Husen 1977) and Indonesia (Nasoetion et al. 1976) suggest that teachers who have received in-service training are more effective than those without such upgrading. According to the Indonesian study, a new curriculum or new textbooks are likely to be more effective if teachers are retrained. In this case, some teachers attended special courses to prepare them for the new primary-school textbooks, and the subsequent performance of pupils and teachers demonstrated that there was significant interaction between the two measures (Husen, Saha, and Noonan 1978).

The importance of in-service training for teachers has been clearly established in several other developing countries as well, but little experimental research has been carried out on the relative costs and effectiveness of different types of training. McMeekin (1982) suggests that cost-effectiveness could in principle be used to evaluate alternative proposals for improving teacher training; for example, the costs of short vacation courses for primary-school teachers combined with radio and correspondence courses could be compared with the costs of building a new teacher training college to provide two-year full-time courses.

Such choices are commonplace in developing countries, which usually have to weigh many arguments for and against each option. On the one hand, in-service training in existing colleges can produce upgraded teachers relatively cheaply and quickly since there is no capital expenditure and no time lag. On the other hand, full-time two-year courses would produce teachers with far more specialized knowledge and higher educational qualifications than those of untrained teachers. Cost-effectiveness analysis does not prove which of these alternatives is better. Rather, it is simply a means of quantifying such arguments in terms of the effect on costs and output indicators.

Of course, the analysis would have to cover many other factors, one of which would be the likely dropout from the two types of training. In many developing countries, a significant proportion of graduates from teacher training colleges either do not become teachers or leave after a short period to become clerks or to enter other occupations. The upgraded teachers, meanwhile, may be more likely to remain in the teaching profession.

Other factors to consider would be the relative salaries of fully trained and upgraded teachers and demographic trends, which might favor the creation of a long-term capacity in the form of a new training college or the greater flexibility of in-service training. Experience suggests that reducing the wastage of qualified teachers may be more cost-effective than increasing the enrollment in teacher-training colleges.

The point of these examples is that cost-effectiveness analysis may be helpful in evaluating alternatives—whether they are connected with educational technology, methods of vocational education, or teacher training. It may also provide guidance in designing projects, as well as in choosing between them. The methodological descriptions of cost-effectiveness analysis (McMeekin 1982; Levin 1983; McMeekin and Gittinger 1984) and the examples of how it has been applied in developing countries provide few lessons for improving educational efficiency. This is not their purpose. The value of cost-effectiveness analysis, it must be reiterated, is not that it proves how best to train teachers or whether to introduce new technology; rather, it helps to focus attention on the factors that determine the relative efficiency of the alternatives. As McMeekin and Gittinger (1984, p. 6) have explained, cost-effectiveness analysis "does not give 'right' or unarguable answers to questions about expenditure choices. It gives, instead, a systematic framework for gathering and analyzing information upon which to base better-informed decisions." One recent example of the use of such a framework is a World Bank evaluation of a diversified secondary-school curriculum.

The Cost-Effectiveness of Curriculum Reform

A theme that runs through many of the proposals for improving the quality and efficiency of education in developing countries is the need to reform the curriculum. This step, it is argued, will equip children with basic skills, improve vocational and technical preparation in schools, widen student choice, increase the employability of school leavers, and reduce wastage and repetition by making the curriculum more relevant to local needs. Curriculum reform may also reduce excess demand for secondary or postsecondary education by providing valued skills and credentials at the end of primary or secondary schooling. Many countries have attempted to diversify the secondary-school curriculum by placing greater emphasis on vocational preparation and skills and by integrating prevocational education more closely with the academic curriculum; the intention has been to raise the status of prevocational courses and to produce schools that are neither purely academic nor purely vocational, but that offer a balanced mixture of general and vocational education. This belief in the merits of a diversified secondary-school curriculum has guided World Bank lending for education projects for many years. Between 1963 and 1982, 117 World Bank education projects contained a diversified curriculum component, and 20 percent of all education sector lending over the past twenty years can be attributed to diversified education components, at a cost of about US$800 million at 1983 prices. The effectiveness of this strategy has recently been examined in the Diversified Secondary Curriculum Study (see chapter 3), which evaluates curriculum diversification in Colombia and Tanzania with respect to both external and internal efficiency (Psacharopoulos and Loxley forthcoming).

The broad objectives of curriculum diversification include meeting manpower demand, rectifying academic bias to make the curriculum more relevant to the needs of the labor market, and improving the overall quality of secondary education and the employment of school leavers. Thus, the objectives of diversification relate to external as well as internal efficiency. Many curriculum development projects have had difficulty implementing changes owing to the lack of specialized teachers and instructional materials and inadequate teacher training.

These problems can reduce the internal efficiency of diversified secondary schools or reduce the effectiveness of curriculum reform. Preliminary evaluations of projects involving curriculum diversification found little evidence to suggest that the new type of schools improved the quality of education, changed students' attitudes toward the labor market, or had the intended effect on employment prospects. The relative overall efficiency of the projects, as measured by lower unit costs and lower repetition rates, was difficult to assess, however, since no in-depth evaluation had ever been attempted. To evaluate the internal and external efficiency of a diversified curriculum, the World Bank, in cooperation with governments, therefore decided to undertake detailed case studies in two countries (Colombia and Tanzania) where diversification had taken place long ago and was reasonably well implemented (Psacharopoulos and Loxley forthcoming). The method chosen was longitudinal tracer studies of students at three points in time: while they were in school, one year after graduation, and several years after graduation.

The impact of curriculum diversification on internal efficiency was measured by cognitive achievement and attitudes to modernity; the test scores of pupils following a diversified curriculum were then compared with pupils in traditional schools. In Colombia, pupils in diversified schools called INEMs were compared with those in traditional (control) schools. Cognitive achievement was tested in academic subjects and in five specialized "tracks" of vocational subjects: commercial, industrial, agricultural, pedagogical, and social service. When the effects of school characteristics and family background were taken into account, it was found that despite considerable differences between subject areas, the following general patterns were discernible:

- Students concentrating in a particular curriculum track always scored higher on tests of their own specialization than did others not specializing in that subject.
- When controls for differences in ability and socioeconomic background were introduced, INEM students always scored significantly higher in both vocational tests and academic subjects than did non-INEM students.
- INEM schools have successfully raised the average level of achievement, particularly in vocational subjects, without sacrificing academic subjects.

When these gains in achievement are compared with the costs, after differences in the ability and family background of students have been taken into account, differences emerge between the specialized tracks: the academic and commercial tracks cost more in INEM schools, but the extra cost is matched by higher average achievement scores; industrial and agricultural tracks cost considerably less in INEM schools and also achieve better results in both academic and vocational subjects; social service tracks cost less in INEM schools, and also achieve better vocational and academic scores (see table 8-5). Thus, the new diversified schools are clearly more cost-effective in the agricultural, industrial, and social service tracks, while in the case of commercial and academic tracks, gains in average test scores are achieved at the expense of extra cost per pupil. The final conclusion is that INEM schools do boost pupil achievement, though in some subjects they also have higher costs. In some cases, therefore, curriculum diversification can increase the internal efficiency of secondary education.

In Tanzania, secondary schools are highly selective, and the aim of curriculum diversification is to prepare students for specific employment or training. All lower secondary schools are supposed to follow a diver-

Table 8-5. *Unit Cost and Achievement by School Type and Curriculum, Colombia, 1981 Cohort*

Curriculum	Unit cost (pesos)	INEM incremental cost (percent)	Academic achievement[a]		Vocational achievement[a]	
			Test score	INEM incremental advantage	Test score	INEM incremental advantage
Academic control	22,200		50			
Academic INEM	25,700	+16	53	+3	—	—
Commercial control	23,200		49		56	
Commercial INEM	25,200	+9	49	+0	61	+5
Industrial control	31,900		47		46	
Industrial INEM	25,000	−21	55	+8	62	+16
Agricultural control	33,700		48		58	
Agricultural INEM	26,200	−22	49	+1	62	+4
Social service control	27,800		48		53	
Social service INEM	25,000	−10	48	+0	57	+4

— Not applicable.

a. Adjusted for differences in ability and socioeconomic background of students.

Source: Psacharopoulos and Loxley (forthcoming).

sified curriculum in which pupils choose between an academic stream and four vocational subjects: commercial, agricultural, technical, or domestic science. Having chosen their vocational subject, pupils take special courses in this area as well as in academic subjects. The study in Tanzania therefore compared average test scores across vocational streams to determine whether vocational specialization boosted pupil achievement in vocational skills. In Tanzania, vocational and academic streams were compared (see figure 8-4), rather than diversified and control schools, as in Colombia. In both countries, pupils perform better in their own special subject area, but because out-of-school factors, such as family background or ability, do not appear to determine who is selected for different curriculum streams, any differences in achievement are presumably the effect of school variables rather than background variables. The question is, what extra cost is associated with the improvement in achievement?

When all the data on unit costs and achievement scores are assembled (table 8-6), we find that all prevocational streams involve extra costs. The unit costs of the most expensive vocational stream, agriculture, were nearly 20 percent higher than those of the academic stream; for this extra cost, subject specialization boosted achievement in agriculture by only two points and in mathematics by only one point, and in English there

Figure 8-4. *Academic and Vocational Test Scores, by Curriculum Track, Tanzania, 1981 Cohort*

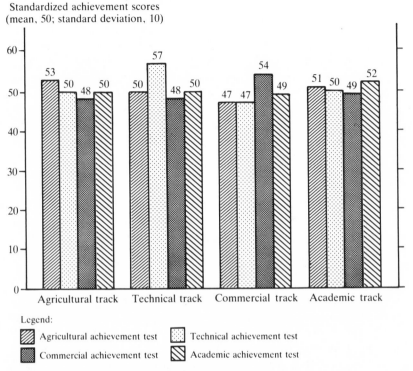

Standardized achievement scores
(mean, 50; standard deviation, 10)

Legend:

▨ Agricultural achievement test   ⬚ Technical achievement test

▨ Commercial achievement test   ◩ Academic achievement test

Source: Psacharopoulos and Loxley (forthcoming).

Table 8-6. *Recurrent Costs and Achievement Scores in Public Schools, Tanzania, 1981 Cohort*

Stream	Unit cost (shillings)	Percentage difference from control group	Incremental advantage in achievement score				
			Math-ematics	En-glish	Agri-culture	Tech-nical	Com-merce
Agriculture	3,449	+19	+1	0	+2	—	—
Technical	3,263	+13	+4	-1	—	+6	—
Commerce	3,160	+9	-2	-1	—	—	+5
Academic (control)	2,888	—	—	—	—	—	—

— Not applicable.
Source: Psacharopoulos and Loxley (forthcoming).

was no gain. The costs of the technical stream were 13 percent higher than the academic stream, and students achieved much better results in technical subjects and mathematics. For commerce, a cost difference of 9 percent meant a five-point advantage in commerce scores, but a decline in mathematics and English scores. In this case, the improvement in vocational skills was achieved at the cost of academic skills. Thus, the general conclusion is that technical specialization yields a substantial increase in vocational skills, but requires more resources per pupil; agriculture costs nearly 20 percent more and increases vocational skills, but not academic skills; and in commerce there is a tradeoff between vocational and academic skills. The technical stream is therefore most cost-effective. In the case of agriculture, a fairly modest gain in achievement requires substantial extra cost, while for commerce, vocational gains must be balanced against academic losses.

Clearly, the decision to introduce a diversified secondary curriculum cannot be justified solely on the basis of internal efficiency. External efficiency, particularly with regard to the effect of vocational preparation on employability and earnings, must also be considered. Since only a small number of school leavers had entered the labor market at the time of the case studies, however, external efficiency cannot be adequately assessed in Colombia and Tanzania. Even though these indicators may be tentative, they are nonetheless important since the main rationale for diversified secondary schooling and the introduction of vocational and prevocational courses has been to increase the economic relevance of schooling. (The implications of comparing the costs and financial benefits of diversified vocational curricula with traditional academic courses were discussed more fully in chapter 3.) Although the data on the earnings of graduates in Colombia and Tanzania are limited, it appears that the diversified secondary curriculum may not offer substantial cost-benefit advantages over the traditional secondary curriculum. Both cost-benefit and cost-effectiveness comparisons suggest external and internal efficiency do improve somewhat, but the yield from the investment in curriculum reform does not yet appear to be very great.

This conclusion should not be interpreted as an attack on the diversified secondary curriculum. Research and evaluation have only begun, and much more information is needed, particularly with respect to the long-term effects of the curriculum. The point is that the results so far have shown that many more questions need to be answered about the relative cost-effectiveness of different patterns and combinations of school organization and curriculum. Increased costs cannot be automatically translated into increased quality and effectiveness. As the 1980 Education Sector Policy Paper pointed out, although curriculum development has usually been given a high priority in educational reform, its effects "have often failed to meet expectations" (World Bank 1980, p. 33). Therefore, any attempt to reform the curriculum must be accompanied by careful evaluation of the effects on educational outputs and outcomes and the effects on costs. Otherwise it will be impossible to identify the most cost-effective ways of changing the curriculum.

Examinations and Selection Procedures

Many of the measures of output and quality that we have discussed are based on examination results and scores. They may be used directly as measures of achievement, or they may be used indirectly to determine other indicators of output; for example, the proportion of pupils who repeat a grade or who drop out of school before completing an educational cycle may be influenced by the selection examinations that determine chances of continuing to the next stage. In particular, examinations that determine pupils' chances of proceeding from primary to secondary education, or that determine which of various types of institution or stream they will enter, will have a decisive effect on measures of output from primary school and often cause high rates of wastage and repetition in developing countries.

There is universal agreement that examinations may be inefficient for selection purposes, since they may fail to measure skills, knowledge, and ability accurately or may fail to predict future levels of achievement. Furthermore, they may distort the curriculum and teaching methods as both teachers and pupils become dominated by the examination and lose sight of wider educational objectives. This attitude—which some have called the diploma disease—can be seen in many developing countries (Dore 1976). Numerous critics have also pointed to the pernicious and capricious results of selection examinations. Yet many also agree that in conditions of limited resources and buoyant private demand for education, some method of selection is inevitable, and in those circumstances examinations perform a useful function.

Several countries have attempted to make their examinations systems more efficient and more equitable. A notable example is Kenya, where the certificate of primary education (CPE) was reformed in 1976 to make it more relevant to the needs of society and more effective in raising the quality of education in Kenya. A review of this experience points to some interesting lessons about the effects of examination reform (Somerset 1982).

Kenya's highly selective CPE examination is used to select pupils for secondary schools, which contain sufficient places for only about 13 percent of primary-school leavers. The examination also serves as a

terminal examination for the majority. In Tanzania, the proportion able
to continue to secondary education is much lower, yet the government
has attempted to reduce reliance on written examinations and to use
continuous assessment as a basis for selection. This effort has not yet
been successful at the primary level, however, and there is still heavy
reliance on external assessment through examination results. Somerset
(1982) therefore concludes that selection examinations still have a long
life in front of them in developing countries, many of which face the same
problem as Kenya: how to improve the effectiveness of the examination
system. Kenya's reform was intended to achieve the following goals:

- *Efficiency:* examination questions should discriminate effectively
  between those able to succeed in secondary schools and less able
  pupils.
- *Equity:* questions should not discriminate unfairly against under-
  privileged pupils.
- *Relevance:* questions should test skills and knowledge relevant to the
  needs of society and encourage teachers to develop the skills school
  leavers need for successful employment and self-employment as well
  as entry to secondary school.
- *Overall higher quality:* the examination should provide an incentive
  to teachers to raise the overall quality of basic education.
- *Improved distribution:* the incentives provided by the examination
  should aim to reduce rather than increase geographical and other
  disparities in achievement levels.

The content of examinations and the type of questions asked were
changed in the early 1970s in an attempt to test a broader range of skills,
to give greater emphasis to relevance, and to discriminate more effectively between the most able and less able pupils. For example, more
emphasis is now given to testing analytical skills and less to memory. In
addition, more attention is paid to the content of questions, so that
relevant knowledge in such areas as basic health care, nutrition, or the
effects of soil erosion is tested. Second, an information feedback system
has been developed to provide schools with both incentives and guidance.
Examination results are therefore used to monitor schools' performance
and to provide information that will enable teachers to improve their
pupils' performance. The interactions between these methods and goals
are summarized in figure 8-5.

The reform of CPE in Kenya involved fundamental changes as well as
technical improvements, such as greater standardization of scores, detec-
tion of cheating, and the like. At the same time, an Examinations
Research Unit was set up to monitor the effects of the changes and to
assess the reform. A preliminary evaluation of the reform shows that

Figure 8-5. *Certificate of Primary Education: Modes and Goals of the Reform Program, Kenya*

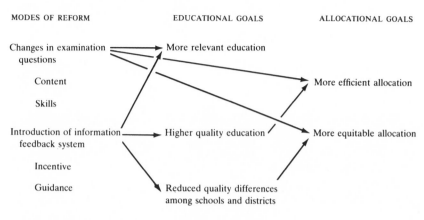

Source: Somerset (1982)

there are often tradeoffs between efficiency and equity, for although questions may be effective in testing relevant skills and efficient in discriminating between children of high and low ability, their content may be biased against pupils in low-cost schools. As Somerset (1982, p. 101) has noted, "The core conflict in these questions is thus between equity and relevance. In this, the questions reflect conditions in the wider society."

Analysis of the effects of the information feedback system has also revealed tradeoffs between efficiency and equity goals. The original purpose of providing schools with detailed information about examination performance was to provide incentives to raise quality and reduce disparities between high- and low-quality schools. Instead, the initial effect of giving schools detailed information about their pupils' examination scores was to increase rather than to reduce differences in quality. The most effective schools with respect to examination successes were able to respond more quickly to the information and thus were able to maintain or even increase their advantage. Later, however, the less successful schools were able to adapt their performance, and the inequalities in examination scores were reduced. At the outset of the program in 1976, the standard deviation of scores increased, but after 1979 it fell, when districts that had been lagging behind showed striking performance gains (table 8-7). This reflects the efforts being made to provide teachers in low-cost schools with the guidance they need to teach these skills more effectively.

Table 8-7. *Standard Deviations of Mean Total Standard Scores, in Certificate of Primary Education, Kenya, 1976–81*

District	1976	1977	1978	1979	1980	1981
All	9.84	11.57	12.54	13.09	11.79	10.15
Rural only	8.32	9.04	9.35	9.36	8.14	7.14

Source: Somerset (1982).

Kenya's experiment provides an interesting example of how the reform of examinations may contribute to both greater efficiency and equity. It also shows, however, that the relationships between the content and design of examinations and pupil performance are often complex, and that conflicts may arise between the effectiveness of examinations in discriminating between able and less able pupils, the relevance of content, and fairness to disadvantaged pupils.

The Use of Capital Resources

Because teacher salaries and other recurrent costs account for a large share of the total cost of education, their effects on internal efficiency have been widely studied, whereas the potential for using buildings and other capital resources more efficiently has sometimes been neglected. Several developing countries, particularly in Latin America, have experimented with double-shift schools, however, in an attempt to use existing buildings and equipment more intensively. The extent to which more intensive use of physical facilities can increase efficiency obviously depends not only on methods of school organization, but also on the availability of teachers and the spare capacity, which is considerable in some developing countries. In Morocco, for example, almost all the schools have excess space, and a survey in Ghana showed excess capacity of more then 28 percent in general classrooms, 43 percent in special-purpose classrooms, and 33 percent in science laboratories (Coombs and Hallak, 1972, p. 208).

A number of World Bank projects have attempted to improve the utilization of existing facilities, for example, through the increased use of double-shift systems (as in Colombia, El Salvador, and Indonesia), the rotating use of classrooms (as in Cameroon, Guyana, and Sierra Leone), and the use of school buildings for evening programs for adults (as in Colombia, El Salvador, Guyana, and Tanzania). Some problems have arisen, however, in introducing rotating use of classrooms in Sierra Leone and Cameroon, and the double-shift system has been difficult to implement in several countries.

Considerable attention is paid to the utilization of capital resources provided in Bank-financed projects, several of which have run into problems of underutilization, particularly with respect to laboratories, workshops, and other specialized facilities. In many cases, these areas have not been fully utilized because specialist teachers have been in short supply, although in some cases enrollment has been lower than planned. Nevertheless, experience in Bank projects shows that more intensive use of facilities can improve internal efficiency in many developing countries by allowing enrollments to increase without introducing additional capital expenditure. Furthermore, the efficiency of school construction can be improved considerably through better design and detailed analysis of building costs (Woodhall 1972).

The Effects of Preschool Investments

Finally, investment in preschool facilities is sometimes proposed as a way of increasing the internal efficiency of education because, as research in the United States has pointed out, "the character of a school's output depends largely on a single input, namely the characteristics of the entering children" (Jencks 1972, p. 256). This argument has been used to justify increased preschool intervention through compensatory programs. Furthermore, the link demonstrated in several developing countries between malnutrition and poor performance in school or dropout and repetition has led to demands for increased investment in nutrition supplements for preschool children as a way of improving their achievement in primary school (Selowsky 1976). Recent reviews (Smilansky 1979; Grawe 1979) have cast doubt on whether such an investment should have a high priority where the goal is educational efficiency. Also, analysis is lacking on the effects of preschool education in the later learning and earning process (Psacharopoulos 1982). Needless to say, research is needed on preschool programs in developing countries, perhaps through collaboration between the World Bank and UNICEF. There is certainly evidence, however, that preschool investment may improve health and nutrition, and that the interrelationships between health and education are important (see chapter 10).

References

Alexander, Leigh, and John Simmons. 1975. *The Determinants of School Achievement in Developing Countries: The Educational Production Function.* World Bank Staff Working Paper no. 201. Washington, D.C.

Arriagada, Anna-Maria. 1983. Determinants of Sixth-Grade Student Achievement in Peru. Washington, D.C.: World Bank, Education Department.

Averch, Harvey A., Stephen Carroll, Theodore S. Donaldson, Herbert J. Kiesling, and John Pincus. 1974. *How Effective Is Schooling? A Critical Review of Research*. Rand Educational Policy Study. Englewood Cliffs, N.J.: Educational Technology Publications.

Beebout, H. 1972. The Production Surface for Academic Achievement: An Economic Study of Malaysian Secondary Schools. Ph.D. diss., University of Wisconsin, Madison.

Berstecher, D. 1970. *Costing Educational Wastage: A Pilot Simulation Study*. Paris: Unesco.

Carnoy, Martin, and Hans Thias. 1974. Draft Report of Second Tunisia Education Research Project. World Bank Research Project no. 248. Washington, D.C.

Coleman, James S., Ernest Q. Campbell, Carol J. Hobson, James McPartland, Alexander M. Mood, Frederick D. Weinfall, and Robert L. York. 1966. *Equality of Educational Opportunity*. Washington, D.C.: U.S. Government Printing Office.

Comber, L. C., and John P. Keeves. 1973. *Science Education in Nineteen Countries*. Stockholm: Almqvist and Wiksell.

Coombs, Philip, and Jacques Hallak. 1972. *Managing Educational Costs*. New York: Oxford University Press.

Dore, R. P. 1976. *The Diploma Disease: Education, Qualification and Development*. London: Allen & Unwin.

Futagami, Shigenari. 1981. Marshalling, Managing and Evaluating the Mass Media for Education and Development. In Gloria Feliciano et al., *The Educational Use of Mass Media*. World Bank Staff Working Paper no. 491. Washington, D.C.

Grawe, Roger. 1979. *Ability in Pre-Schoolers, Earnings and Home Environment*. World Bank Staff Working Paper no. 322. Washington, D.C.

Haddad, Wadi. 1978. *Educational Effects of Class Size*. World Bank Staff Working Paper no. 280. Washington, D.C.

——. 1979. *Educational and Economic Effects of Promotion and Repetition Practices*. World Bank Staff Working Paper no. 319. Washington, D.C.

Hartley, M. J., and E. V. Swanson. 1984. Achievement and Wastage: An Analysis of the Retention of Basic Skills in Primary Education. Draft monograph. Washington, D.C.: World Bank, Development Research Department.

Heyneman, Stephen P. 1980. Differences between Developed and Developing Countries: Comment on Simmons and Alexander's "Determinants of School Achievement," *Economic Development and Cultural Change* 28, no. 2 (January):403–06.

Heyneman, S., and W. Loxley. 1983. The Effect of Primary School Quality on Academic Achievement across 29 High- and Low-Income Countries. *American Journal of Sociology* 88, no. 6 (May), pp. 1162–94.

Heyneman, Stephen, Joseph Farrell, and Manuel Sepulveda-Stuardo. 1981. Textbooks and Achievement in Developing Countries: What We Know. *Journal of Curriculum Studies* 13(3):227–46.

Heyneman, Stephen, Dean T. Jamison, and Xenia Montenegro. 1984. Textbooks in the Philippines: Evaluation of the Pedagogical Impact of a Nationwide Investment. *Educational Evaluation and Policy Analysis* 6(2):139–50.

Husen, Torsten. 1967. *International Study of Achievement in Mathematics: A Comparison of Twelve Countries*. Vols. 1 and 2. New York: Wiley; and Stockholm: Almqvist and Wiksell.

———. 1977. Pupils, Teachers and Schools in Botswana: A National Evaluative Survey of Primary and Secondary Education. In *Education for Kagisono*. Report of the National Commission on Education in Botswana. Gaborone: Government Printing Office.

Husen, Torsten, Lawrence Saha, and Richard Noonan. 1978. *Teacher Training and Student Achievement in Less Developed Countries*. World Bank Staff Working Paper no. 310. Washington, D.C.

International Bureau of Education (IBE). 1971. *Wastage in Education: A World Problem*. Paris: Unesco/IBE.

———. 1972. *A Statistical Study of Wastage at School*. Paris: Unesco/IBE.

Jamison, Dean, and François Orivel. 1982. The Cost-Effectiveness of Distance Teaching for School Equivalency. In *Alternative Routes to Formal Education*, ed. Hilary Perraton. Baltimore, Md.: Johns Hopkins University Press.

Jamison, D. T., Barbara Searle, Klaus Golda, and Stephen Heyneman. 1981. Improving Elementary Mathematics Education in Nicaragua: An Experimental Study of the Impact of Textbooks and Radio on Achievement. *Journal of Educational Psychology* 73 (August):556–67.

Jencks, Christopher. 1972. *Inequality: A Reassessment of the Effect of Family and Schooling in America*. New York: Basic Books.

Lee, Kye-Woo. 1981. Equity and an Alternative Educational Method: A Korean Case Study. *Comparative Education Review* 25, no. 1 (February):45–63.

Lee, Kye-Woo, Shigenari Futagami, and Bernard Braithwaite. 1982. The Korean Air-Correspondence High School. In *Alternative Routes to Formal Education*, ed. Hilary Perraton. Baltimore, Md.: Johns Hopkins University Press.

Levin, H. M. 1983. *Cost-Effectiveness: A Primer*. Beverly Hills, Calif.: Sage.

McAnany, Emile. 1980. In Unesco, *The Economics of New Educational Media*. Vol 2. Paris: Unesco.

McMeekin, Robert. 1975. *Educational Planning and Expenditure Decisions in Developing Countries, with a Malaysian Case Study*. New York: Praeger.

———. 1982. Cost-Effectiveness Analysis of Alternative Teacher Training Projects. World Bank Economic Development Institute Case Study and Exercise Series EC-027. Washington, D.C.

McMeekin, Robert, and J. P. Gittinger. 1984. Cost-Effectiveness Comparison of Vocational Education Alternatives. World Bank Economic Development Institute Exercise Series 135/018. Washington, D.C.

Nasoetion, N., A. Djalil, I. Musa, and S. Soelistyo. 1976. *The Development of Education Evaluation Models in Indonesia*. Paris: Unesco/International Institute for Education Planning.

Neumann, Peter. 1980. *Publishing for Schools: Textbooks and the Less Developed Countries*. World Bank Staff Working Paper no. 398. Washington, D.C.

Neumann, Peter, and Maureen Cunningham. 1982. *Mexico's Free Textbooks: Nationalism and the Urgency to Educate*. World Bank Staff Working Paper no. 541. Washington, D.C.

Nollen, Stanley. 1975. The Economics of Education: Research Results and Needs. *Teachers' College Record* 77, no. 1 (September):51–77.

Peaker, Gilbert F. 1975. *An Empirical Study of Education in Twenty-one Countries: A Technical Report*. Stockholm: Almqvist and Wiksell.

Perraton, Hilary, ed. 1982. *Alternative Routes to Formal Education*. Baltimore, Md.: Johns Hopkins University Press.

Psacharopoulos, G. 1982. The Economics of Early Childhood Education and Day-Care. *International Review of Education* 28 (1):53–70.

Psacharopoulos, G., and William Loxley. Forthcoming. *Diversified Secondary Education and Development: Evidence from Colombia and Tanzania*. Baltimore, Md.: Johns Hopkins University Press.

Schiefelbein, E. 1975. Repeating: An Overlooked Problem of Latin American Education. *Comparative Education Review* 19 (October):468–87.

Schiefelbein, E., and J. Simmons. 1978. *The Determinants of School Achievement: A Review of the Research for Developing Countries*. IDRC Document MR-9. Ottawa: International Development Research Centre.

Schiefelbein, E., J. P. Farrell, and M. Sepulveda-Stuardo. 1983. *The Influence of School Resources in Chile: Their Effect on Educational Achievement and Occupational Attainment*. World Bank Staff Working Paper no. 530. Washington, D.C.

Selowsky, Marcelo. 1976. A Note on Pre-School-Age Investment in Human Capital in Developing Countries. *Economic Development and Cultural Change* 24, no. 2 (July):707–19.

Shuluka, S. 1974. Achievements of Indian Children in Mother Tongue (Hindi) and Science. *Comparative Education Review* 18, no. 2 (June):237–47.

Smilansky, Moshe. 1979. *Priorities in Education: Pre-School; Evidence and Conclusions*. Bank Staff Working Paper no. 323. Washington, D.C.

Somerset, H. C. A. 1982. Examinations Reform: The Kenya Experience. Washington, D.C.: World Bank, Education Department.

Thias, Hans, and Martin Carnoy. 1969. *Cost-Benefit Analysis in Education: A Case Study of Kenya*. Baltimore, Md.: Johns Hopkins University Press.

Thorndike, Robert L. 1973. *Reading Comprehension in Fifteen Countries*. International Studies in Evaluation. New York: Halsted Press.

Unesco. 1977. *Development of School Enrollment: World and Regional Statistical Trends and Projections 1960–2000*. International Conference on Education. Paris.

――――. 1982. *The Economics of New Educational Media*. Vol. 3. *Cost and Effectiveness Overview and Synthesis*. Paris.

U.S. Department of Health, Education, and Welfare (HEW). 1970. *Do Teachers Make a Difference? A Report on Recent Research on Pupil Achievement*. Washington, D.C.: Government Printing Office.

Vaizey, John, with Keith Norris, John Sheehan, et al. 1972. *The Political Economy of Education*. London: Duckworth.

Woodhall, Maureen. 1972. The Use of Cost Analysis to Improve the Efficiency of School Building in England and Wales. In *Educational Cost Analysis in Action: Case Studies for Planners*. Vol. 3. Paris: International Institute for Educational Planning.

World Bank. 1974. *Education*. Sector Policy Paper. Washington, D.C.

――――. 1980. *Education*. Sector Policy Paper. Washington, D.C.

Zymelman, Manuel. 1973. *Financing and Efficiency in Education*. Harvard University Press.

9

Equity Considerations in Educational Investment

Selecting an investment project raises both efficiency and equity issues. Decisions must be based on a number of criteria, some of which have been discussed in the preceding chapters. We now turn to equity, which concerns the way the costs and the benefits of an investment are distributed among different groups in society. With educational investment, the question is whether the costs and benefits are equally distributed among regions, and whether males and females, and different social, economic, or ethnic groups, have equal access to educational facilities. Because the inequalities of access and distribution are particularly severe in developing countries, the distributional effects of investment projects and the way in which they are financed are important in these countries. This chapter looks at some of the wider issues concerning the equity of educational investment.

Four questions in particular need to be examined if we are to determine the effects of educational investment on equity: How are educational resources and facilities distributed among different areas or groups? What are the effects of government subsidies for education on the distribution of costs and benefits and the distribution of total income or welfare? Can educational investment be used to redistribute wealth, income, and opportunities between rich and poor, advantaged and disadvantaged? How effective is education as a redistributive tool? Such questions are now given high priority in developing countries and also in the World Bank and other international agencies, whereas in the past questions about the efficient allocation of resources tended to dominate the selection of investment projects. Moreover, equity is now likely to be an explicit criterion in the assessment and evaluation of education projects.

With the increasing interest in equity issues, more information is becoming available on the equity implications of educational investment. One question that cannot be simply resolved, however, is whether equity and efficiency are mutually consistent goals and criteria, or whether they conflict. In some cases, equity and efficiency criteria appear to point to the same conclusion. Because of the great differences in urban and rural participation, for example, expanding education in rural areas is likely to serve equity objectives and also promote efficiency by encouraging the more rapid adoption of improved agricultural methods. Similarly, greater equity would be achieved by increasing educational access for females; with a better-qualified female population, national development is likely to be fostered through the changes that can be expected in the nature of labor force participation and through gains in family welfare, family planning, and health and child care.

The evidence on rates of return to education (see chapter 3), which shows high rates of return to primary schooling, taken together with evidence on the redistributive effects of public expenditure on education (illustrated in table 9-1) suggests that in some developing countries expanding primary education not only would be a profitable investment, but would promote equity, since in general primary education tends to redistribute resources toward the poor. Furthermore, as primary education becomes more widely disseminated, additional spending will be increasingly concentrated on backward rural areas and disadvantaged groups.

Thus the *World Development Report* of 1980 and 1981 suggested that efficiency and equity goals could be jointly served by educational investment, and the 1980 Education Sector Policy Paper concluded that equity in education and economic development goals are mutually consistent (World Bank 1980a, b, 1981). It would indeed be convenient for educational investment strategy if efficiency and equity criteria always pointed in the same direction. Unfortunately, this is not the case. Although many education projects do contribute to both efficiency and equity goals, many others find that the criteria fail to agree. As a result, many countries

Table 9-1. *Public Education Spending per Household,*
by Income Group
(dollars)

Income group[a]	Malaysia, 1974[b]		Colombia, 1974[c]	
	Primary	Postsecondary	Primary	University
Poorest 20 percent	135	4	48	1
Richest 20 percent	45	63	9	46

a. Households ranked by income per person.
b. Federal costs per household.
c. Subsidies per household.
Source: World Bank (1980b), p. 50.

have found themselves in an "equity-efficiency quandary" (Schultz 1972), which has become of increasing concern to governments as public pressure for more schooling has come into conflict with budgetary pressures.

Behrman and Birdsall (1983a), for example, have challenged the view that school investments permit harmonious pursuit of equity and efficiency goals. Estimates of the returns to improving school quality in Brazil led them to argue that there is an equity-productivity tradeoff and that the conventional wisdom about investment in education in developing countries may promote substantial overinvestment of resources in schooling and direct funds toward the wrong items.

Thus, many disagree about the relationship between efficiency and equity criteria and about the relative weights to be attached to efficiency and equity goals. This chapter does not attempt to resolve the controversy. Its purpose is to examine the research on equity implications of educational investment and the conclusions about equity in assessments of World Bank projects. First, however, we should consider what is meant by the term *equity*, how it can be measured, and whether there are alternative ways of measuring the distribution of education.

Opinions vary as to the meaning of equity because the concept is closely bound up with the notion of fairness. As a result, differences inevitably arise in judgments of what is a fair or equitable distribution of wealth, income, or education. There are also different ways of measuring the distribution. Any discussion of equity implications must therefore incorporate both analytical techniques and interpretative judgment. In examining the analytical techniques that are used to explore equity implications of educational investment, we show why the controversy concerning the contribution to equity goals persists, and we examine the tradeoffs between equity and efficiency. That is to say, we look at positive issues in this chapter. We do not, however, address the normative issue of whether equity or efficiency goals should take precedence, because this is a matter of political judgment. The question should nevertheless be considered when investment projects are being selected, particularly if the goals conflict. Research on equity issues cannot resolve this normative question, but it can inform politicians and policymakers about the likely results of their choices and about possible tradeoffs between them.

The Concept of Equity

A distinction must be made between normative and positive statements at the outset of any discussion of equity. As noted earlier, equity not only refers to the distribution or sharing of resources among individuals or groups, but it is also tied up with the notion of justice. Any determination of equity must therefore be based on facts about how resources are distributed and on normative judgments about how society should distribute resources. Judgments will differ in this regard because societies differ in their moral or philosophical principles. Even the factual aspect of equity analysis will involve judgments, however, about how groups to which resources will be distributed should be differentiated. That is to say, the analysis of how goods are distributed cannot proceed unless the population is first classified into mutually exclusive groups. The basis for classification may be age, sex, social class, income level, occupation, or any other relevant variable.

There is ample evidence that some groups in developing countries have better access to education than others, but that the factors determining access vary among countries. One review (Fields 1980) finds considerable differences in educational participation of individuals classified by sex, socioeconomic background, urban and rural areas, and also race, language, and religion. In Malaysia, for example, disparities exist not only between males and females and between different geographical regions, but also between those of Malay and Chinese origin. In Sri Lanka ethnic and religious differences play a role in educational attainment, while in Peru language is a determining factor (table 9-2). Income level is a significant factor in Colombia and Malaysia (table 9-3), and in a number of countries enrollment rates vary between urban and rural areas (table 9-4). Conclusions about the distribution of education therefore depend on how the population is classified and also on how educational participation is measured.

Similarly, the terms of measurement must be clearly defined where income distribution is concerned. If monetary income is used, the value of subsistence income, which is often important in developing countries, will be overlooked. If individual income is used as a measure, the results will not be the same as when total household income is used. Furthermore, weekly, annual, or lifetime income estimates will all lead to different conclusions about the distribution of income. Most of the research on education and income distribution has focused on earnings, but wages and salaries represent only 50 percent of national income in many developing countries, and the distribution of earned income differs markedly from the distribution of unearned income.

The point is that the unit of measurement may affect the analysis of distributional issues, as can be seen from the divergent conclusions reached by two studies of the distribution of educational costs and benefits in Colombia, one of which classified families by total household income (Jallade 1974) and the other by per capita income (Selowsky 1979). It is not that one measure is correct and one incorrect. Both are

Table 9-2. *Educational Attainment by Race, Tribe, Language,
or Religious Group, Selected Developing Countries*

Country and group	Educational attainment (percent)				
	No formal schooling	Primary school	Some or completed secondary	Post-secondary	Not specified
Malaysia, 1974					
Males, Kuala Lumpur					
Malay	2.0	42.7	48.7	6.6	—
Chinese	3.5	43.5	47.9	5.2	—
Males, East Coast towns					
Malay	9.4	55.2	33.4	2.0	—
Chinese	4.1	40.0	52.4	3.8	—
Females, Kuala Lumpur					
Malay	1.5	25.7	66.2	6.6	—
Chinese	7.1	39.0	51.6	2.3	—
Females, East Coast towns					
Malay	14.8	10.9	50.6	3.7	—
Chinese	7.7	30.9	57.7	1.9	—
Peru, 1972					
Language					
Spanish	21.6	54.5	19.1	3.9	1.0
Quechua	50.3	43.7	4.8	0.7	0.5
Aymará	42.2	51.7	5.0	0.6	0.4
Other autochthonous	56.9	35.7	3.1	0.6	3.7
Foreign	6.2	27.8	37.5	20.9	7.6
Total	30.2	51.2	14.8	3.0	0.9

Sri Lanka, 1967	Population	University admission
Ethnic group		
Sinhalese	71.0	84.1
Ceylon Tamil	11.1	14.1
Muslim	6.7	1.4
Burgher	0.4	0.1
Indian Tamil	10.6	0.1
Other	0.2	0.2
Total	100.0	100.0
Religion		
Buddhist	66.3	79.2
Hindu	18.4	10.9
Muslim	6.9	1.4
Christian	8.3	7.8
Other	0.1	0.7
Total	100.0	100.0

— Not applicable.
Source: Fields (1980), p. 261.

Table 9-3. *Enrollment Rates by Income Level, Colombia and Malaysia*

		Enrollment ratios (percent)		
Country	Income level[a]	Primary	Sec-ondary	All levels
Colombia, 1970 (Urban)	0–6,000	n.a.	n.a.	48.5
	6,000–12,000	n.a.	n.a.	43.1
	12,000–18,000	n.a.	n.a.	45.4
	18,000–24,000	n.a.	n.a.	48.2
	24,000–30,000	n.a.	n.a.	48.7
	30,000–36,000	n.a.	n.a.	47.6
	36,000–48,000	n.a.	n.a.	52.0
	48,000–60,000	n.a.	n.a.	54.4
	60,000–72,000	n.a.	n.a.	52.9
	72,000–84,000	n.a.	n.a.	56.8
	84,000–120,000	n.a.	n.a.	58.3
	120,000–180,000	n.a.	n.a.	64.9
	180,000–240,000	n.a.	n.a.	66.2
	Over 240,000	n.a.	n.a.	61.9
	Total			50.8
Colombia, 1974	1	72.2	17.1	
	2	84.0	21.9	
	3	86.6	28.2	
	4	95.1	43.8	
	5	89.8	62.5	
	Average	82.9	31.5	
Malaysia, 1974	1	85	n.a.	
	2	86	n.a.	
	3	93	n.a.	
	4	99	n.a.	
	5	90	n.a.	
	Mean	90	n.a.	

n.a. Not available.

a. For Colombia, 1970, the income level is in pesos per year; for Colombia, 1974, and Malaysia, it is quintiles of household per capita income (poorest to richest).

Source: Fields (1980), p. 263.

perfectly valid measures of family income, but it is a matter of judgment as to which is the more appropriate for assessing equality or inequality of distribution.

When it comes to normative judgments about fairness or justice, we can expect even more disagreement about relevant criteria. Furthermore, according to McMahon (1982), at least three types of equity can be discerned in the considerable literature on this subject: horizontal equity, which is usually taken to mean equal treatment of equals; vertical equity,

Table 9-4. *Distribution of Population by Educational Attainment in Urban and Rural Areas, Selected Developing Countries*
(percent)

Country	Year	Area	No schooling	Pri-mary	Sec-ondary	Post-secondary
Algeria	1971	Total	84.4	13.0	2.2	0.3
		Urban	73.5	20.5	5.2	0.8
		Rural	89.9	9.2	0.6	0.1
Chile	1970	Total	13.1	61.0	22.2	3.8
		Urban	8.3	60.1	27.0	4.8
		Rural	29.8	64.2	5.4	0.6
Colombia	1973	Total	22.4	55.9	18.4	3.3
		Urban	14.2	54.8	26.1	4.9
		Rural	38.4	58.0	3.5	0.2
Dominican Republic	1970	Total	40.1	45.9	12.1	1.9
		Urban	22.9	49.5	23.5	4.1
		Rural	52.8	43.2	3.7	0.3
Guatemala	1973	Total	93.9		4.9	1.2
		Urban	85.2		11.8	2.9
		Rural	98.7		1.1	0.2
India	1971	Total	72.2	22.7	3.9	1.1
		Urban	46.6	36.8	12.3	4.2
		Rural	78.6	19.2	1.8	0.3

Country	Year					
Indonesia	1971	Total	55.3	39.1	5.1	0.5
		Urban	22.0	56.9	14.1	7.0
		Rural	45.2	51.4	2.1	1.3
Kenya	1969	Total	75.9	20.5	2.8	0.8
		Urban	46.9	37.0	5.3	10.7
		Rural	79.5	18.5	1.1	0.9
Korea, Rep. of	1970	Total	72.6 ⎱		21.8	5.6
		Urban	53.0 ⎰		36.0	11.0
		Rural	86.2		12.0	1.8
Malaysia, West	1970	Total	40.6	44.6	9.3	5.5
		Urban	32.2	42.7	14.0	11.1
		Rural	43.9	45.4	7.4	3.2
Paraguay	1972	Total	19.6	68.0	10.3	2.1
		Urban	11.2	63.6	20.4	4.8
		Rural	25.5	71.1	3.1	0.2
Sri Lanka	1971	Total	29.5	58.9	9.4	2.3
		Urban	20.6	58.8	16.8	3.9
		Rural	32.3	58.9	7.1	1.7
Tunisia	1966	Total	89.1	7.1	3.0	0.7
		Urban	78.4	13.7	6.3	1.7
		Rural	96.4	2.7	0.8	0.1
Yemen, People's Dem. Rep. of	1973	Total	72.9	22.1		5.1 ⎱
		Urban	59.1	30.2		10.7
		Rural	80.0	17.9		2.1 ⎰

Source: Fields (1980), p. 262.

which refers to unequal treatment of unequals (and which raises the question of how equality or inequality is to be judged); and intergenerational equity, which lies between the other two types of equity and is concerned with ensuring that inequalities in one generation are not simply perpetuated.

Debates about the equity implications of subsidies for higher education often depend on whether a populist view or an elitist (meritocratic) view of equity is adopted (Bowman, Millot, and Schiefelbein 1984). The former would distribute education opportunities equally among members of society, whereas the latter would base the distribution of higher education on ability and thus could be said to promote unequal treatment of unequals. Others might argue that the same could be said of those who discriminate in favor of the poorest or most disadvantaged members of society. Their definition of equity, however, might lead to conclusions that would conflict with those suggested by a meritocratic definition.

The terms *equity* and *equality* are obviously far from synonymous, although they are sometimes used interchangeably, particularly in discussions about the distribution of educational opportunities. Many of the analytical tools designed to measure dispersion and distribution compare an actual distribution with a theoretical distribution of perfect equality. But this does not mean that a perfectly equal distribution is necessarily the desired or favored objective of government policy on education or income distribution. According to some definitions, as we have seen, the desired goal is unequal treatment of unequals. Thus when concepts of equity are being debated, the issue is usually a philosophical rather than an economic one (Alexander 1982). This means that an economic analysis of the equity implications of investment decisions must begin with a clear statement of what is judged to be an equitable distribution of resources, or at least a clear definition of the criteria for judging equity.

A distinction must also be made between access to education and participation, because the distribution of enrollments may be related to the numerous factors that influence private demand (see chapter 5) as well as to the access to education in different regions or geographical areas. That is to say, government policy in some countries may be concerned with reducing inequalities of access and thus may opt for building schools in remote areas or for reducing fees to remove financial barriers for those who cannot afford to enroll. In other countries, the overriding concern may be to reduce inequalities of participation, and thus incentives may be provided for those who do not choose to enroll. In this case, the policy chosen to promote equity will be wider than one that is simply concerned with removing barriers, because it seeks to increase participation by changing some of the factors that govern private demand. Such an approach to equity is likely to involve greater costs than the former. Whether a government adopts a wider or more restrictive approach to questions of equity will depend partly on available resources.

The 1980 Education Sector Policy Paper suggested that the relative concern for access, equity, and efficiency is likely to be a function of the level of educational development in a country. It also observed that when enrollment rates are low (less than 30 percent), governments are likely to be primarily concerned with increasing access to the system by having more schools enroll more students. As enrollment rates grow to more than 70 to 80 percent, the main concern becomes to maximize internal efficiency and ensure equality in the distribution of resources. The paper noted, however, that "efforts to provide a more equitable access to education should not diminish, because the last 5 to 10 percent of students remaining to be enrolled will be the most difficult to serve and will probably require special measures" (World Bank 1980a, p. 29). For that reason, the cost of extending access to the last 10 percent will be high.

Still another distinction should be made between the quantity and quality of educational provision. Most analysis of the distribution of education in developing countries is mainly concerned with quantitative indicators. As Behrman and Birdsall (1983b) show, however, regional or urban-rural differences in the quality of schooling may be substantial, as in the case of Brazil.

Measures of Equality or Inequality of Distribution

Distribution is usually analyzed by first showing the frequency of occurrence of different values of a variable among different groups in a population. A frequency distribution (often expressed in percentages) can be illustrated by means of a frequency curve or histogram and described in terms of its central or average value, such as the median (the value of the variable that divides the distribution into two equal parts), the mean (which is obtained by dividing the total of all values by the number of observations), or the mode (the maximum or peak of the frequency distribution).

A frequency distribution can also be described in terms of quartiles (which divide the distribution into four equal parts), quintiles (fifths), deciles (tenths), or percentiles. These measures can be used to compare the relative share going to specific groups, such as the top decile, or the bottom quintile. Measures of dispersion can also be used to analyze frequency distributions. These include standard deviation, coefficient of variation, and measures of skewness, which indicate whether observations tend to be concentrated above or below the mean.

More precise measures of equality or inequality of distribution are

Table 9-5. *Distribution of Primary Education among Provinces in a Hypothetical Country*

Prov- ince[a]	Total provincial populations		School-age populations		Enrollments		Cumu- lative school-age popu- lations (percent)	Cumu- lative enroll- ments (percent)	Enroll- ment ratio (percent)	Repre- sentation index
	Thousands	Percent	Thousands	Percent	Thousands	Percent				
C	400.0	33.8	80	40	20	20	40	20	25	50
A	111.1	9.4	20	10	10	10	50	30	50	100
B	312.5	26.5	50	25	30	30	75	60	60	120
D	357.1	30.2	50	25	40	40	100	100	80	160
Total	1,180.7	100	200	100	100	100	—	—	—	—
Average	—		—		—		—	—	50	100
Gini coefficient = 0.245										

a. Note that provinces have been ordered according to their respective enrollment ratios (beginning with the lowest) to facilitate graphing and subsequent calculation.

Source: ter Weele et al. (1979).

needed, however, for the detailed analysis of the distribution of income
or wealth in a society, or the distribution of educational opportunities.
The most commonly used measures of distribution are the representation
index, the curve of concentration (generally known as the Lorenz curve),
and the Gini coefficient. In all these cases the distribution of income or
the provision of education is compared with a perfectly equal distribution. In other words, the actual share of a group is compared with the
amount that group would have received if all groups in society received
equal shares.

The representation index measures the equality or inequality of distribution on the basis of the relative shares of different groups. The index
is a simple device that indicates whether a particular group or area is over-
or underrepresented in relation to the total population. Table 9-5 and
figure 9-1 illustrate the unequal geographical distribution of enrollment in
a hypothetical country with four provinces. The representation index is

Figure 9-1. *Representation Index for Primary-School Enrollment*
in Four Provinces

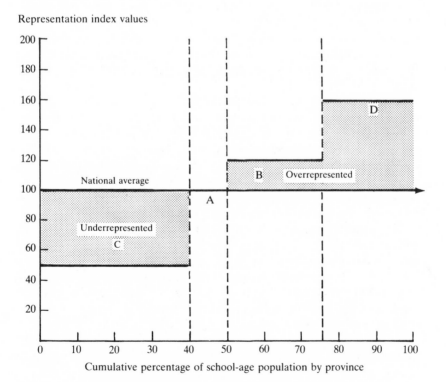

Source: ter Weele et al. (1979).

calculated by dividing the share of enrollments in each province by the share of the school-age population. The geographical distribution in this case appears inequitable since two provinces are overrepresented (their index is more than 100) and one is underrepresented (its index is less than 100).

This is only one way to judge geographical equity, however. Other criteria might yield different conclusions. Suppose that instead of looking at the distribution of enrollment, we looked at the distribution of school facilities. The conclusion about equity might then be reversed. In some developing countries a region that appears to enjoy a relative advantage where enrollment rates are concerned may appear to be disadvantaged when the availability of school facilities is considered. In this case, the advantage suggested by the enrollment figures is the result of overcrowding. This example shows that judgments about the equality of distribution of education depend on the criterion chosen for measuring the distribution of benefits.

Such judgments become even more difficult to make if we wish to take into account the distribution of both costs and benefits. A region or province that has high enrollment rates, for example, may also provide a high proportion of the country's total tax revenue. If its share of school enrollments is compared with its share of the school-age population, it may appear to be overrepresented (that is, the index has a value of more than 100). If tax revenues are regarded as a measure of the province's contribution to educational costs, however, then its share of costs may exceed its share of benefits, and it would appear disadvantaged. This situation would be considered inequitable if it is assumed that a region should receive a share of benefits roughly equal to its share of taxes. This assumption is open to question, of course, and once again raises the question of what should be considered equitable. Since there is no unambiguous definition of an equitable distribution of resources, the equity implications of alternative investment proposals must be analyzed by means of alternative measures and definitions.

The Lorenz curve, also called the curve of concentration, is another way of comparing the relative shares of different regions, groups, or individuals. If all groups receive an equal share, then 10 percent of the population will receive 10 percent of the total resources, 20 percent will receive 20 percent of the resources, and so on. In this case, the Lorenz curve would be drawn as a diagonal straight line. Any deviation from this diagonal indicates inequality.

A Lorenz curve can be calculated for the distribution of school enrollments shown in table 9-5, which also shows the cumulative share of school-age population and school enrollments. Figure 9-2 shows the curve of concentration, or Lorenz curve, in which the cumulative shares of enrollments are plotted against the cumulative shares of the school-age population. The Lorenz curve deviates from the diagonal, and the extent of this deviation can be seen in the shaded area in figure 9-2. The shaded area is therefore a measure of the degree of inequality in the distribution. The Gini coefficient provides a precise measure of inequality by expressing the shaded area as a ratio of the total area below the diagonal. The

Figure 9-2. *Curve of Concentration for Cumulative Primary Enrollments in Relation to School-Age Population*

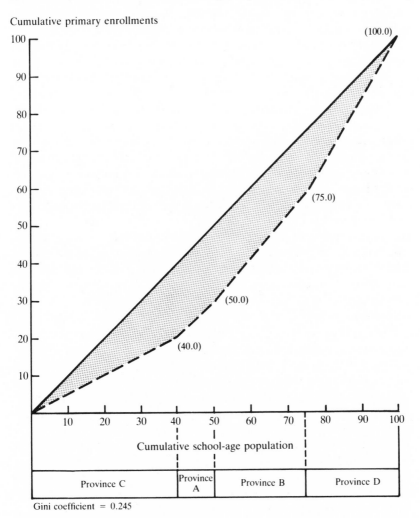

Gini coefficient = 0.245

Source: ter Weele et al. (1979), p. 8.

value of a Gini coefficient always lies between 0 (where there is perfect equality of distribution and no deviation from the diagonal) and 1.0 (which represents the hypothetical case of complete inequality, where one group or area monopolizes all scarce resources). The higher the value of the Gini coefficient, the greater the degree of inequality.

All these measures of inequality of distribution depend on the way the population is divided into groups for purposes of analysis. In the example in figures 9-1 and 9-2, province D appears to be at a relative advantage since it contains 40 percent of total enrollments and only 25 percent of the total school-age population. Closer analysis of school enrollment statistics may reveal, however, that the distribution within the province is quite uneven and that urban and rural areas vary considerably. In such a case, parts of the province may enjoy virtually 100 percent enrollment, while rural areas may have highly unequal distribution. Under these circumstances, any conclusions about the geographical equity of resource allocation in the country and any recommendations about how to allocate new investment in order to improve equity will have to take into account the distribution within provinces as well as between provinces.

The same problems arise in an analysis of the distribution of resources between different groups in society. Tools such as the representation index, Lorenz curve, and Gini coefficient do not, by themselves, demonstrate that policies are equitable or inequitable, but they may help to show how resources are distributed in a society, whether a particular investment will lead to greater or less geographical inequality, and whether it will favor advantaged or disadvantaged groups (for example, by indicating whether it will tend to redistribute income or resources from the rich to the poor or vice versa). The following sections provide some examples of such analysis.

The Geographical Distribution of Primary-School Enrollments

Considerable inequalities in the provision of education exist in many developing countries. Resources are unequally distributed between countries, between urban and rural areas within countries, between males and females, and between different social groups. It would be possible to reduce these inequalities by allocating extra resources to the most relatively deprived countries, areas, or groups, provided that these could be accurately identified and provided that governments and lending agencies such as the World Bank gave high priority to the redistribution of resources. Leaving aside, for the moment, the political question of whether extra resources can, and should, be earmarked for the most deprived sections of the population, we should ask whether it is possible to identify these areas of relative deprivation.

The World Bank developed a computer program, the school location-allocation planning system, to analyze the degree of inequality of distribution of primary-school enrollments in fifteen East African countries and to investigate whether selective allocation of educational investment could help reduce these inequalities (Maas and Criel 1982). With the aid of this program, researchers have found that the degree of inequality in the distribution of educational opportunities varies enormously from one East African country to another. In some countries, educational distribution is fairly even, whereas in other countries enrollments or the allocation of school places varies greatly among districts, between urban and rural areas, and between the sexes (see table 9-6 for the representation index for each country and figure 9-3 for the Lorenz curve for the region as a whole). The Gini coefficients for total enrollment in individual countries (figure 9-4) range from a low of 0.06 for Swaziland (which indicates considerable equality of distribution) to 0.558 in Southern Sudan (which indicates a high degree of inequality).

An interesting correlation was revealed in this study between Gini coefficients, which measure relative inequalities, and national enrollment rates, which provide a measure of the absolute level of provision in the country as a whole. The strong inverse relationship between the two indices is shown clearly in figure 9-4. The study therefore concludes that countries with relatively high national participation rates tend to have a fairly equal distribution of enrollments (that is, a low Gini coefficient), and countries with relatively low national enrollments have a relatively less equal distribution of enrollments (that is, a high Gini coefficient). In other words, according to Maas and Criel (1982, p. 9), "The degree of *relative* inequity in the allocation of educational opportunities is negatively correlated to the *absolute* level of provision of the social service. Both elements are, in fact, mutually reinforcing."

This finding has important implications for decisions to invest in new educational facilities since it suggests that if a country succeeds in raising the average level of enrollment, overall inequalities may be reduced. The precise effect, however, will depend on how the incremental resources are allocated. The school location-allocation planning system has been used to simulate the redistributive effects of a planned increase in enrollments on the assumption that the incremental resources are allocated to the poorest or most deprived areas through what is described as a "bump-up" technique.

Before a "bump-up" can be applied, the representation index of each district must be calculated separately and all districts ranked in relation to this index. Resources are then allocated to the most disadvantaged

Table 9-6. *Basic Education Data for Fifteen Countries, East Africa*

Country	Enrollment	School-age population	Repre-sentation index	Enrollment rate (percent)	Primary places funded in bank projects to fiscal 1981	
					Number	Cost (US$ million)
Botswana	156,295	162,661	132.3	96.086	0	0
Burundi	145,234	622,000	32.2	23.350	27,000	9.0
Ethiopia	1,379,653	4,200,436	45.2	32.845	110,000	8.1
Kenya	3,698,170	2,968,280	171.6	124.590	4,550	6.0
Lesotho	226,316	198,254	157.2	114.155	7,500	3.5
Madagascar	1,389,936	1,196,000	161.1	116.968	0	0
Malawi	704,549	1,237,206	78.4	56.947	59,200	8.5
Rwanda	401,521	606,000	91.3	66.258	6,000	8.0
Somalia	263,749	585,000	62.1	45.085	0	0
Sudan	1,015,269	2,532,588	55.2	40.088	12,000	2.6
Southern Sudan	106,805	580,893	25.3	18.386	9,000	1.8
Swaziland	105,607	111,580	130.4	94.647	11,500	4.4
Tanzania	2,913,222	3,213,433	124.9	90.655	24,000	11.7
Uganda	1,314,363	1,913,378	94.6	68.693	0	0
Zaire	3,293,000	3,798,000	119.4	86.704	2,400	2.1
Zambia	936,724	999,027	129.1	93.764	7,580	6.0
Total	18,250,648	25,136,941	100.0	72.604	280,730	71.7

Note: The fiscal year of the World Bank runs from July 1 to June 30.
Source: Maas and Criel (1982), p. 27.

district (that is, the one with the lowest representation index) to bump it up until the value of the index reaches that of the second lowest point. The exercise is repeated until the additional resources are exhausted. At this point, a revised Gini coefficient and representation index are calculated in order to measure the effect of the additional investment on the inequality of distribution. This type of exercise can be used to explore the following questions:

Figure 9-3. *Lorenz Curve for Total Primary Enrollment in East Africa*

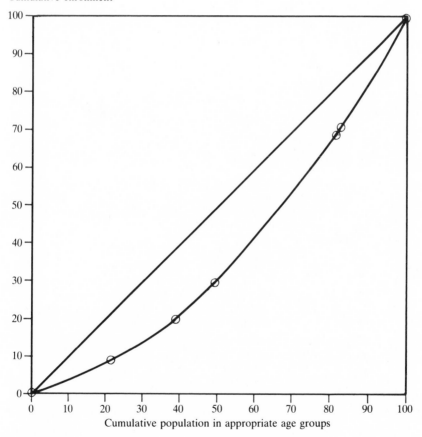

Cumulative enrollment

Cumulative population in appropriate age groups

Legend:
o = total enrollment

Source: Maas and Criel (1982).

Figure 9-4. *Relationship between Unequal Distribution
and Enrollment Rates, East Africa*

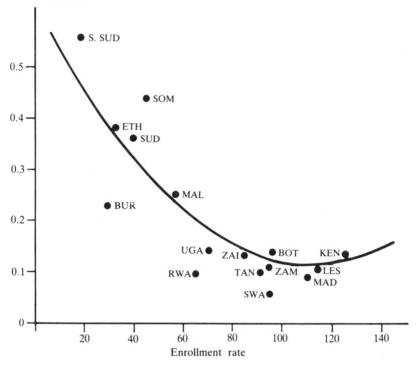

Note: See table 9-7 for the full names of countries.
Source: Maas and Criel (1982).

- If the resources could be found, where should they be invested so as
  to have the greatest impact on the most deprived areas and reduce
  the inequalities within and between countries?
- What would the cost implications of such an investment be?
- In which countries and in which areas should investments be made?

The effect of various increases in enrollments in East Africa, which
ranged from 1 percent (182,506 new school places) to 10 percent
(1,825,064 new school places) was estimated on the assumption that all
the investment was concentrated in the most disadvantaged countries or
districts. According to the bump-up method of allocation, an investment
of 182,506 school places would have the greatest impact on regional
inequalities if it were concentrated in Burundi, Ethiopia, and Southern

Sudan. If enough resources were available to provide a 10 percent increase in enrollments, however, the 1,825,064 new school places should be distributed between Burundi, Ethiopia, Somalia, Sudan, and Southern Sudan, but such a policy would have high costs.

This exercise showed which countries would have highest priority if the primary objective of educational investment was to reduce geographical inequalities. Analysis of male-female inequalities (see table 9-7) or urban-rural inequalities would present a different picture, in which case a different order of priorities might be justified if the objective of investment was to reduce inequalities between sexes or between city dwellers and the rural population.

Such an analysis cannot, of course, provide a simple answer to the question of where investment should be allocated. There are many reasons—including demographic, historical, geographic, and economic ones—why certain areas are relatively badly endowed with school facilities. The final conclusion of this exercise is that, for a given level of expenditure, there is often a tradeoff between maximizing school places (which probably means adding to already advantaged areas) and minimizing overall inequalities (which would mean higher unit costs). Ulti-

Table 9-7. *Male-Female Representation Levels in Fifteen Countries*

	Male		Female	
Country	*Representation index*	*Enrollment rate (percent)*	*Representation index*	*Enrollment rate (percent)*
Botswana	88.7	85.2	111.4	107.0
Burundi	121.1	30.9	78.8	20.1
Ethiopia	133.0	43.7	65.5	21.5
Kenya	105.8	131.8	94.2	117.4
Lesotho	83.0	95.0	116.4	133.0
Madagascar	106.3	116.9	93.9	103.3
Malawi	123.5	70.3	78.1	44.5
Rwanda	105.0	67.2	95.1	60.9
Somalia	122.9	55.4	75.1	33.9
Sudan	123.2	49.4	77.9	31.2
Southern Sudan	138.9	25.5	55.2	10.1
Swaziland	101.3	95.9	98.7	93.5
Tanzania	108.1	98.0	91.8	83.3
Uganda	113.2	77.7	86.3	59.3
Zaire	118.9	100.6	80.3	68.0
Zambia	106.8	100.1	93.1	87.3

Source: Maas and Criel (1982), p. 15.

mately, this tradeoff has to be considered a political issue that cannot be readily resolved (Maas and Criel 1982, p. 9).

Although an analysis of geographical inequities cannot, by itself, suggest where investments should be allocated, it can shed some light on the equity implications of alternative policies. When the priorities suggested by this analysis are compared with the actual distribution of investment in primary-school places financed through the World Bank, some of these resources appear to have been allocated to the most disadvantaged countries (Ethiopia, Burundi, and Sudan). At the same time, other countries that were shown to be underrepresented in terms of enrollments (for example, Somalia) have benefited much less from World Bank lending than countries that exhibit much less inequality, such as Tanzania or Swaziland.

Improving equity is not, of course, the only objective of World Bank lending, and this analysis does not suggest that it should be. Maas and Criel (1982, p. 33) conclude, however, that "at a time when the promotion of equitable participation in development is going to have to compete increasingly with other lending priorities, it would be appropriate to include these findings and targets in the formulation of an overall development program for the education sector." The geographical distribution of resources is, of course, only one dimension of equity.

The Effect of Educational Investment
on Income Distribution

Like the distribution of education, the distribution of income, earnings, or wealth in a society can be analyzed by means of Lorenz curves, Gini coefficients, and other measures of inequality such as the coefficient of variation (see Atkinson 1970; Sen 1973). Indeed, economists were concerned with the analysis of income distribution long before the effects of the distribution of education received much attention. The World Bank has commissioned research in this area for many years. Among recent examples are a joint study by the World Bank and the Economic Commission for Latin America (ECLA) on income distribution in Colombia, Chile, Honduras, Mexico, Panama, and Peru (Altimir 1982). A similar project has been undertaken by the Bank and the Economic and Social Commission for Asia and the Pacific (ESCAP) in twelve countries in Asia.

Many practical problems arise in measuring income distribution, and here, too, different methods of measurement produce different conclusions. Estimates of income distribution based on household income in Latin America, for example, differ from estimates derived from national income. Recent calculations for Argentina, Brazil, Colombia, Chile, Mexico, and Uruguay suggest that income distribution is more unequal when household income data are used as the basis for measurement (Altimir and Sourouille 1980). Other calculations (Carnoy 1979) have shown that conclusions about changes in the inequality of earnings distribution in Mexico depend on how inequality is measured because different measures give different weight to those with very high incomes.

Whatever measures are used, wide disparities will be found in the distribution of income and wealth, particularly in developing countries. A general trend in the relation between income inequality and economic development was first identified in the 1950s by Simon Kuznets (1955, 1956), who demonstrated that the incomes of the poorest 40 percent of the population normally grow more slowly than the average until reaching a critical point (in figure 9-5 at approximately $800 per capita), after which they begin to grow faster than the average. Thus, income inequality first increases, then decreases in the course of economic development.

The *World Development Report* for 1980 emphasized, however, that the Kuznets curve is not an iron law. Although a clear relationship exists between income inequality and level of development (see figure 9-5), some countries lie well above the curve and others below. Evidence of changes in income inequality in particular developing countries confirms that they do not all follow a path of the same shape. The report therefore

Figure 9-5. *Income of Poorest Groups*

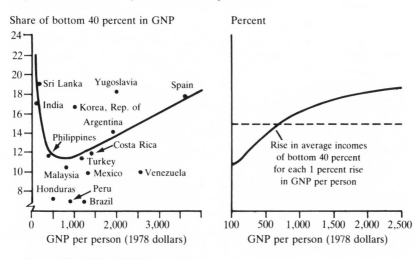

Source: World Bank (1980b), p. 40.

concluded that "much depends on government policy, which can reduce the unevenness of the modernization process—and accelerate growth—by promoting productivity gains in traditional small-scale agriculture, increasing the rate at which labor is absorbed into the modern industrial sector, and not concentrating public investment and services on a few places and social groups" (World Bank 1980b, p. 41).

This conclusion has bearing on educational investment for two reasons. First, if education can promote productivity gains in agriculture and help labor become absorbed into the modern industrial sector, then educational investment may help to reduce income inequality. Second, the distribution of educational opportunities may be used as a tool to redistribute income. This means ensuring that investment and services are not concentrated on a few favored places and social groups, and that educational investment is consciously used to redistribute income and raise the incomes of the poor.

Since education is a strong determinant of earnings, investments that help to equalize educational opportunities may help to equalize incomes in the future. The role of education in reducing inequalities of income distribution has therefore attracted considerable attention. Early work in this area concentrated on the relationship between education and income distribution in the United States and other developed countries (Becker and Chiswick 1966). During the 1960s, the development of human capital theory engendered considerable optimism about the effects of educational investment. Not only did economists argue that investment in education would contribute to economic growth and development, but they saw it as a powerful, long-term method of equalizing earnings and income distribution.

This earlier optimism has recently been replaced by pessimism, which in the United States can be explained by the fact that there has been little or no reduction in the inequality of income after three decades of rapid educational expansion. The persistence of income inequality has been attributed to various factors: the influence of social class and family origins (Bowles 1972); cyclical unemployment (Chiswick and Mincer 1972); job competition (Thurow 1975); and even pure chance (Jencks 1972). At the same time, research in the United Kingdom suggests that equalizing the distribution of schooling may help to equalize the distribution of income within age groups, even though it may make the distribution of income more unequal among the population as a whole (Blaug, Dougherty, and Psacharopoulos 1982).

An increase in the provision of education may, under certain circumstances, increase some aspects of income inequality while reducing other inequalities. The effectiveness of educational investment as a redistributive tool has therefore come under considerable scrutiny in both developed and developing countries. In addition, both the International Labour Organisation and the World Bank have commissioned research on the links between education, employment, and income distribution and the redistributive impact of education investment and government subsidies. The general conclusion of these various studies appears to be that educational investment may affect the distribution of income in developing countries in the following ways:

- Education may raise the overall level of income and thus reduce the absolute level of poverty.
- It may change the dispersion of income.
- It may open up new opportunities for the children of the poor and thus act as a vehicle for social mobility.
- Alternatively, if participation is confined to the children of the rich, education may simply transmit intergenerational inequality.
- If certain groups (for example males, city dwellers, or certain ethnic groups) obtain higher financial rewards from their education than other groups (such as females, inhabitants of rural areas, and ethnic minorities), then education may increase income inequalities.
- The pattern of financing, in particular the extent of public subsidies for education, may redistribute income from those who are taxed to those who benefit from subsidies.
- Education may interact with fertility, mortality, health, and other aspects of development that affect income distribution.

These effects can be examined by means of earnings functions, which analyze the relationship between various independent variables—including age, education, family background, and occupation—and the dependent variable, earned income. Another technique is to use decomposition studies to identify the contribution of the various factors that determine income differences. One review of this research concluded that "education is the single most important determinant of income. That is, if you sought to determine an individual's income and could ask only one question, you would do best to ascertain how much education the individual in question has received" (Fields, 1980, p. 238).

The explanatory power of education has also been explored in various studies by means of path analysis, which is a technique for analyzing the influence of various factors on a dependent variable. The underlying assumption is that causal relationships follow a certain pattern. If, for example, earnings are assumed to be influenced by four variables—family background, intellectual ability, education, and occupation—the chronological sequence could be expressed by a simple path model:

Family background → Intellectual ability → Education → Occupation → Earnings.

The actual causal relationships are far more complex than this, however. Thus path analysis usually consists of establishing the correlations and interrelationships between the variables and calculating the direct effects on the dependent variable and the interactions between intermediary variables. Figure 9-6 illustrates a path model used to explore the direct and indirect influence of family background, ability, academic achievement, and attendance at diversified secondary schools on earnings in Colombia (Psacharopoulos and Loxley forthcoming). Such studies provide ample evidence that education is a powerful determinant of both individual and group income differences. Because the benefits of education are greater for some groups (for example, urban males) than for others, the way in which educational opportunities are distributed in society will affect future income distribution. As we have just pointed out, however, the causal relationships between education and income inequalities are complex and need further research.

When countries rather than individuals or groups are compared, the

Figure 9-6. *Path Model, Colombia—Academic Achievement as Intermediate Variable*

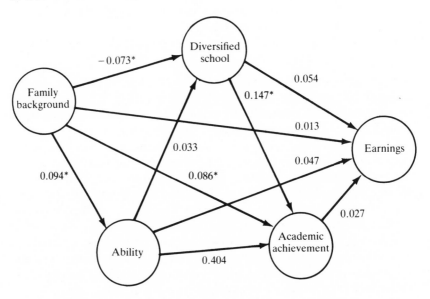

$R^2 = 0.007$
$N = 510$
*Statistically significant path 0.10 level
Source: Psacharopoulos and Loxley (forthcoming)

relationship between education and income inequality becomes even clearer. As might be expected, given the relationship between level of economic development and income inequality noted above, the higher the level of school enrollments in a country, the less the income inequality (Adelman and Morris 1973; Chenery and Syrquin 1975; Ahluwalia 1976). Evidence from comparisons of forty-nine developed and developing countries (Psacharopoulos 1978; Winegarden 1979) also indicates that the greater the inequalities of educational attainment of the population, the greater the income inequality in a country. Does this mean that a reduction in educational disparities will necessarily lead to a more equal income distribution? The evidence on this point is inconclusive.

A number of studies in Latin America, for example, have reached conflicting conclusions. When Carnoy (1979, p. 98) examined empirical studies of education and earnings distribution in Brazil, Chile, Cuba, Mexico, and Peru, he found a paradox:

> Schooling apparently plays a very important role in determining individual earnings in Latin America, but the distribution of education in the labor force is not very important in influencing earnings distribution. Rather, government incomes policy, affecting the reward to different levels of schooling, different work sectors, different types of occupations and different regions of the country, may be a much more important factor in understanding changes in income distribution.

Carnoy also suggested that although the distribution of schooling in Mexico has become more equal, income distribution has become more unequal, whereas Marin and Psacharopoulos (1976) suggested that expanding primary education in Mexico would make income distribution more equal.

Other studies (for example, Jallade 1974) have also drawn attention to the paradox that rapid growth of education in developing countries has not led to a reduction in income inequality in those countries, and point out that in some cases income inequality may even have increased. After reviewing evidence from five developing countries (Brazil, Costa Rica, India, the Philippines, and Sri Lanka), Fields (1978 and 1980) concluded that there is a "closer relationship between educational performance and *aggregate* economic growth than between educational performance and *distribution* (in terms either of relative inequality or of absolute poverty)" (1980, p. 283).

If, in some circumstances, educational investment actually increases income inequality, what are the equity implications? Even this question is surrounded by controversy, as can be seen in two studies (Langoni 1972; Fishlow 1973) that investigated the increased inequality of income in

Brazil during the 1960s. Both agreed that education contributes to increased inequality, but disagreed when it came to equity. Langoni argued that the increased inequality has been a natural but temporary result of the fast growth of the Brazilian economy, especially at the end of the 1960s, whereas Fishlow argued that government policies, which favored a select few, were the main cause of increased income inequality.

Research on the causes and implications of income inequality is continuing in Brazil and in other countries, both developed and developing. Although the relationship between education and income distribution is not yet fully understood, the evidence thus far seems to support the following conclusions: those countries with higher levels of education exhibit less income inequality on average; greater disparities in educational attainments among the population are associated with greater income inequality; and expansion of the lower levels of education is likely to compress the dispersion of earnings and reduce income inequality.

It follows that educational investment designed to reduce educational inequalities is a necessary, but not sufficient, condition for greater equality of income distribution. Carnoy (1979, p. 98) concluded that "educational policy can only contribute to the more equal distribution of earnings when it is carried out in concert with an incomes policy." Redistributive taxation policies will also have a more direct impact on income distribution than education, which works more slowly and indirectly. Just as economic growth may initially lead to greater income inequality—as shown by the Kuznets curve—so educational investment may in some circumstances lead to greater inequalities of income distribution. Nevertheless, an increase in the level of education is likely to contribute eventually to greater equality of income distribution. Thus, even if no immediate impact on income distribution can be expected, policies designed to reduce inequalities of educational opportunity are likely to lead to a more equal distribution of income in the long run. In the short run, however, governments are considering various measures to increase equality of opportunity and make the financing of education more equitable.

The Effectiveness of Measures to Equalize Education

If level of education is correlated with equality of income distribution, we might expect the expansion of educational opportunities to encourage equality by reducing educational disparities. Furthermore, we might expect to find that children from privileged backgrounds tend to monopolize a small school system, while expansion benefits children from less privileged backgrounds. An extensive review of the data on social and income selectivity in education in more than forty countries has concluded, however, that the degree of selectivity in secondary and higher education varies enormously among countries and that only a tenuous association exists between per capita income and the degree of social selectivity in schools: "At each level of per capita income, selectivity varies greatly among societies" (Anderson 1983, p. 235).

A government may wish to reduce social selectivity but not to expand education rapidly on grounds of cost or estimates of manpower demand. Can selection mechanisms, rather than rapid expansion, be used to equalize opportunities and reduce income inequalities? Recent research (Armitage and Sabot 1982) has explored this question in two countries (Kenya and Tanzania) that have adopted different policies on secondary-school enrollments. In examining the equity implications of the two countries' policies, the study has taken intergenerational mobility rather than the extent of inequality as the criterion of equity. Armitage and Sabot assume that even if income is very unevenly distributed, it may still be possible, given appropriate educational policies, for the poor to achieve intergenerational mobility.

Kenya and Tanzania are similar in their level of development, but they differ markedly in the size of their secondary-school enrollments. Kenya has adopted a policy of educational expansion at the secondary level, while Tanzania has restricted the supply of secondary-school places. The question is which of these policies has proved more effective in promoting equality of access and intergenerational mobility. Armitage and Sabot concluded that a policy of expansion is more likely to foster equality of opportunity and intergenerational mobility since, even with meritocratic selection, access to higher education may be confined to socially more privileged students.

According to another study of Kenya and Tanzania (Knight and Sabot 1982), educational expansion in Kenya has reduced the relative earnings differential of secondary-school graduates by roughly 20 percent and therefore has reduced inequalities of income. Thus educational expansion is said to have been an effective means of reducing inequality of wage distribution in this case, even though wages in Kenya have not been under direct government control. Tanzania has adopted the alternative policy of government intervention on wages, but, according to Knight and Sabot, this policy is effective only in the public sector and thus the government's attempt to equalize incomes has been only partly successful. Such research on the relative effectiveness of alternative policies on educational investment and direct intervention in pay policy is still in its infancy in developing countries.

Some economists in developed countries have argued that if governments wish to equalize incomes, they should consider policies designed to redistribute income through progressive taxation or pay policies designed to reduce earnings differentials, which are more effective in this regard than policies designed to redistribute educational opportunities (Jencks 1972). Even though the redistributive impact of education is not as direct and immediate as taxation or pay policy, however, it is still worth taking into account in countries where both income and educational opportunities are unequally distributed.

It is now widely accepted that education will not, by itself, ensure a more equal distribution of income, but this is no justification for undue pessimism about the equity implications of education investment. Blaug (1978, p. 5), for example, argues that although the distribution of schooling as such is not a powerful instrument for equalizing income distribution, particularly in comparison with tax and expenditure policies or with direct interventions in the labor market via income policies, it is nevertheless "one non-negligible instrument for equalising income distribution and it may even be one that is more politically palatable than the more powerful, direct instruments available to governments."

Moreover, in developing countries where inequalities of educational provision are severe, it may be desirable, on equity grounds, to pursue the goal of more equal distribution of educational opportunities even if such action does not have a strong impact on income distribution. However, it is equally important to ensure that the pattern of financing education does not have an undesirable impact on income distribution (see chapter 6).

The Equity Implications of Financing Educational Investment

Whether public subsidies for education redistribute income from the poor to the rich or from the rich to the poor has not been fully established in either developed or developing countries. Some evidence from a number of developing countries suggests, however, that the present pattern of subsidies often favors the rich (Blaug 1982; Fields 1975, 1980; Jallade 1973; Psacharopoulos 1977).

Education is heavily subsidized in developing countries, particularly at the secondary and higher levels. Mingat and Tan (1985) have shown that equity in the distribution of public resources depends on the pattern of subsidization by level of education as well as on the socioeconomic composition of the student population at each level. With respect to the first factor, their study of major world regions reveals that in developing countries the distribution of public resources among members of a given generation of school-age children is strikingly inequitable. For example, in developing countries as a group, 71 percent of the cohort (those with primary or no schooling) share only 22.1 percent of the overall cohort resources, whereas 6.4 percent (those with higher education) get 38.6 percent of these resources. In some regions of the world, the pattern of distribution is even more skewed. In Francophone Africa, for example, 15.7 percent of the resources go to 86 percent of the cohort, whereas 39 percent goes to only 2.4 percent of the cohort who leave the education system with higher education.

When this pattern of distribution in public education resources is overlaid with the socioeconomic composition of the student body, the inequity of providing heavily subsidized secondary and higher education becomes even more apparent. In another study, Mingat and Tan (forthcoming) show that in developing countries other than Francophone Africa, an individual from a nonfarmer home receives 2.5 times as many public education resources as his counterpart from a farming background. In Francophone Africa, the picture is again darker, as the corresponding figure is 3.5 times as many. Within the nonfarmer population, individuals from white-collar backgrounds receive, on the average, roughly five times as much in the way of resources as those from farming backgrounds; in Francophone Africa, the contrast is even sharper, since they receive ten times as much.

Private institutions may receive public subsidies, as in Latin America, and this may have equity implications. These subsidies are financed by and large from general taxation, although in a few cases they are supported by specific taxes for education. Inequalities of access or participation mean that the benefits of education are disproportionately enjoyed by upper-income families, whose children are far more likely to complete secondary schooling or enroll in higher education. Moreover, education increases the earning capacity and thus the lifetime income of the educated, so that those who benefit from education subsidies are likely to have higher-than-average incomes in the future. Therefore, it seems clear that public subsidies of education involve a transfer of income from poor or average-income taxpayers to the children of the rich or to those who may become rich as a result of their education. So runs the argument, and at first sight it appears that education subsidies may indeed redistribute income in the wrong direction from the point of view of equity, that is, from the poor to the rich.

Nevertheless, research on the subject concurs that the equity implications of education subsidies are more complex than this simple argument supposes. Therefore any analysis of the redistributive effects of education subsidies must begin by defining what is meant by an equitable distribution of the financial burden and the benefits of education. Three equity criteria have been suggested: equal opportunity, payment by those who benefit, and ability to pay.

Conclusions about the equity of methods of financing education depend on which criterion is chosen. Fields (1975), for example, considers all three in his analysis of the higher education system in Kenya. If the government's aim is to ensure equality of opportunity, the system is inequitable, since the lowest income group constitutes 90 percent of all taxpayers but provides only 60 to 75 percent of students in higher education, whereas the richest 1 percent of taxpayers provide 6 to 10 percent of students. The fact that all students are heavily subsidized regardless of income also appears to flout the principle "He who benefits should pay." If, however, taxpayers are classified not according to their relative numbers, but to their contribution to total taxation, then the bottom income group provides about 70 percent of taxes and between 60 and 75 percent of students. Thus, the distribution of benefits and financial burdens in Kenya appears to satisfy the criterion that each income group should contribute to the costs of education in proportion to the benefits its children receive, although there is some tendency for the middle-income group to benefit at the expense of the top-income groups.

The third criterion, ability to pay, takes into account unequal incomes and requires those with the greatest ability to pay to make the greatest contribution to costs. In other words, costs and benefits are not distributed proportionately, as they would be under the previous criterion, but payment rises as a function of income. This is not the case in Kenya (table 9-8), and indeed a recent study of secondary education subsidies there concludes that at present the least needy are the ones most likely to obtain a subsidy (Armitage and Sabot forthcoming). It is true in Chile, however (Foxley, Aninat, and Arellano 1976, 1977). As table 9-9 shows, the ratio of costs to benefits increases as income rises. According to the criterion of ability to pay, therefore, educational financing appears equitable in Chile even though the equal opportunity criterion is not met, since the bottom income group constitutes 30 percent of the population but receives less than 15 percent of the benefits (measured in terms of educational expenditure).

This is another example to show that judgments about the equity of educational finance depend on the criteria used. According to table 9-9, the share of education expenditure going to the bottom income groups in Chile is much greater at the primary level than at the university level. The same is true in Colombia, where public subsidies for primary education redistribute income from the rich to the poor, although at the secondary and higher levels the lower- and upper-middle classes are subsidized by

Table 9-8. *Incidence of Taxes and Distribution of Higher Education by Income of Students' Parents, Kenya, 1971*
(percent)

Annual income bracket (shillings)	Distribution of taxpayers	Distribution of taxes	Distribution of students by parents' income class		
			Primary	Secondary	University of Nairobi
0–2,400	90.5	67.9	70.7	74.7	60.2
2,400–3,600	5.4	8.8	3.8	4.0	2.2
3,600–4,800	1.3	2.2	6.2	4.9	2.2
4,800–6,000	0.7	1.4	5.6	4.4	11.8
6,000–8,400	0.5	1.5	6.2	4.7	11.8
8,400–12,000	0.5	2.4	1.9	1.8	2.2
12,000–16,800	...	...	3.4 ⎫	0.9 ⎫	...
16,800–24,000	1.1	15.7	0.8 ⎬ 5.6	2.2 ⎬ 5.5	9.6
More than 24,000	...	...	1.4 ⎭	2.4 ⎭	...
Total	100.0	100.0	100.0	100.0	100.0

... Less than 0.5.
Source: Fields (1975), p. 252.

both the poor and the very rich (Jallade 1974; Selowsky 1979). This finding is reflected in the Lorenz curves for the distribution of income and subsidies for education in Colombia (see figure 9-7). The curve for primary education lies above the diagonal, but the curve for higher education lies well below the diagonal and the income distribution curve. Thus primary-school subsidies are progressive, but higher-education subsidies are highly regressive.

The extent to which the financing of education in developing countries is judged to be equitable thus depends on the criterion used and the way income and educational benefits are measured. It also depends on whether equity is judged from a static or a dynamic point of view. How income is redistributed can be determined from the income levels of today, but today's poor student may become tomorrow's rich taxpayer, perhaps partly as a result of educational investment. Therefore a transfer of income from the rich to the poor today may not reduce income inequalities in the future. In other words, the lifetime distribution of costs and benefits as well as the present balance must be taken into account. Future benefits, however, will depend not only on the extent to which education will raise lifetime income, but also on the extent to which these benefits vary with the quality of education.

Table 9-9. *Distribution of Taxes and Public Expenditures in Education by Income Class, Chile, 1969*
(percent)

Household annual income (minimum annual income units)	Distribution of households (1)	Distribution of direct and indirect tax burdens (2)	Primary (3)	Secondary (4)	University (5)	Other educational expenditure[a] (6)	Total (7)	Cost-benefit ratio (2)/(7) (8)
0–1	29.8	7.6	19.5	11.1	6.2	21.4	14.7	52
1–2	31.6	18.1	38.4	31.0	27.8	36.1	33.8	54
2–3	17.6	17.3	23.6	29.0	16.1	22.0	22.3	78
3–4	7.4	9.6	10.4	12.3	13.8	9.9	11.6	83
4–5	4.5	7.7	3.5	5.6	7.3	3.8	5.0	154
5–6	2.9	7.0	1.2	2.0	4.0	1.4	2.1	333
6–8	2.7	7.8	1.3	3.8	6.4	1.9	3.2	244
8–10	1.5	6.5	0.6	1.5	1.9	0.7	1.1	591
10 or more	2.0	18.2	1.5	3.7	16.5	2.8	6.2	294
Total	100.0	100.0	100.0	100.0	100.0	100.0	100.0	—

a. Includes expenditures on technical and vocational training, provision of free lunch for primary students, other programs of financial assistance, and expenditure on school buildings.
Source: Fields (1980), p. 299.

276

Figure 9-7. *Distribution of Income and of Subsidies for Education, Colombia*

Percent of income or of subsidy

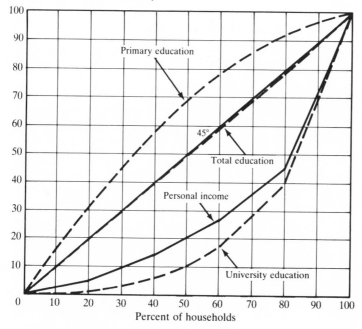

Percent of households

Source: Selowsky (1979), p 24.

Equity and Quality of Education

Most of this chapter has been concerned with disparities in enrollment rates or other quantitative measures of education and with the equity implications of expanding or reducing the quantitative inequalities of access. In developing countries, however, qualitative disparities are often as severe as quantitative differences in provision. Yet much less attention has been paid to the equity implications of qualitative differences. Part of the reason is that in many countries the quantitative disparities are so severe that attention has necessarily focused on reducing these inequalities. Moreover, it is extremely difficult to measure the quality of education.

Some recent research has nonetheless attempted to measure qualitative differences, not only in relation to the costs and benefits and rate of return to educational investment, but also with respect to the equity implications of investment choices. One such study argues on the basis of earnings data for Brazil that failure to control for variations in the quality of schooling may produce biases in the estimation of the returns to schooling (Behrman and Birdsall 1983b). When quality of schooling is measured in terms of the average education of teachers in different regions, the standard deviation is 3.2 years of schooling for teachers, and the social rate of return to investment in improving school quality is probably larger than the social rate of return to investing in school quantity.

This finding has important equity implications. As the study points out, controlling for differences in school quality substantially reduces unexplained regional income differentials. In other words, inequalities of income that cannot be explained solely in terms of differences in the quantity of schooling of male workers appear to be partly due to differences in the quality of schooling. Differences in quality seem to explain not only some of the income differences between individuals, but also regional and urban-rural differences (Behrman and Birdsall 1983b).

When Birdsall (1982) explored differences in quality of schooling in Brazil and their effect on private demand for schooling, she found that the proportion of children who have never attended school varies from 60 to 70 percent in rural areas of the Northeast to less than 10 percent in the urban South (see table 9-10). The average level of schooling of teachers varies from 4.8 years in rural Northeast Brazil to 11.8 years in the urban South (table 9-11). If the length of teachers' education is regarded as a measure of the quality of schooling, then the disparities in quantity of schooling of the children in different regions are reinforced by differences

Table 9-10. *Percentage of Children Aged 8–15 Who Have Never Attended School, by Urban-Rural Residence and Region, Brazil, 1970*

	South		Frontier/ Central		Northeast	
Age	Urban	Rural	Urban	Rural	Urban	Rural
8	10.57	35.77	0	59.26	20.9	74.17
9	5.8	23.42	26.32	51.85	15.0	66.99
10	6.76	17.12	6.67	37.14	17.58	68.37
11	2.28	8.46	0	52.38	14.29	59.18
12	4.59	14.28	10.0	41.66	14.52	65.35
13	2.46	20.61	0	41.67	14.55	61.04
14	3.9	17.86	0	55.0	10.2	54.54
15	4.86	16.67	5.88	57.14	20.76	61.36

Source: Birdsall (1982).

Table 9-11. *Variations in Level of Schooling by Region and Rural or Urban Area, Brazil, 1970*

	South		Central/ Frontier		Northeast	
Characteristic	*Urban*	*Rural*	*Urban*	*Rural*	*Urban*	*Rural*
Children's completed years of schooling	4.48	2.53	4.28	0.82	3.08	0.68
Father's years of schooling	3.75	1.42	4.26	0.85	2.53	0.37
Mean years of schooling of primary- and secondary-school teachers	11.80	7.54	10.10	5.53	10.33	4.79

Source: Birdsall (1982), p. 32.

in quality. Other indications of quality give the same impression. Recurrent expenditure per child is as much as ten times greater in urban than in rural areas. Birdsall suggests that improvements in availability and quality of education in Brazil are of greater benefit to children in poor households and to children with less-educated mothers.

In other developing countries the quality of schooling also varies greatly between regions or between urban and rural locations. Despite the relative lack of systematic research on quality differences, such differences are just as relevant as quantitative disparities in determining how public resources for education should be allocated. The difficulty of financing qualitative improvements in developing countries raises once more the question of the potential contribution of cost recovery mechanisms (see chapter 6), which have been recommended on the grounds of equity as well as efficiency.

Equity Implications of Cost Recovery Mechanisms

As we have seen, high levels of subsidy for education do not necessarily ensure equal opportunities and may even bring about undesirable transfers of income from the poor to the rich. As a result, some have suggested that fees combined with loans constitute a more equitable method of financing education (Rogers 1971; Jallade 1973; Fields 1975; Psacharopoulos 1977; Armitage and Sabot 1984). The argument is that higher-income students would pay a higher proportion of the costs of their education, thus freeing resources for subsidies for the less fortunate, which could take the form of selective scholarships or quantitative expansion and qualitative improvements that would benefit those suffering the greatest inequalities.

As chapter 6 has pointed out, however, a system of student loans does not automatically lead to greater equity since it too may favor upper-income students (Jallade 1974). Unless student loans are linked to an increase in fees, the loans to students in private universities will simply amount to a transfer of income from taxpayers to higher-income students through interest subsidies. Although the introduction of fees and student loans would not, by itself, ensure more equitable distribution of education opportunities, in many cases it would be a step in the right direction and would reduce the anomaly in many developing countries that permits those who reap the greatest financial rewards from education to enjoy the greatest subsidy. A case in point is Ghana, which introduced student loans in 1970 on the grounds of equity. Even though the loan scheme failed because of political problems, the justification for such a policy on grounds of equity remains valid, not only in Ghana but also in other developing countries. As was frequently emphasized during the debate on student loans in Ghana, a system that provides subsidies of N\$3,000 for each university student, compared with only N\$200 for secondary-school pupils and N\$20 for primary-school pupils, offends social justice in a country that cannot afford universal primary education.

The introduction of fees for those who can afford to pay and selective subsidies for the poor would have a positive effect on income distribution in many developing countries. Selective scholarships based on both ability and financial need are obviously one way of targeting subsidies for the poor. Income-related fees are another method, which has already been tried in some universities in Colombia. Such cost-recovery mechanisms cannot, by themselves, reduce inequity and inequality of opportunity, but they may be able to contribute to this end by freeing resources for investments in disadvantaged regions. However, there may be tradeoffs between efficiency and equity objectives.

Conflict and Tradeoffs between Efficiency and Equity

Earlier we asked whether efficiency and equity goals are mutually consistent, or whether they may conflict and thus force policymakers to consider tradeoffs between the two. On a theoretical level, economists have paid considerable attention to the conditions under which equity and efficiency are jointly attainable and to cases in which equity can be increased only by sacrificing efficiency or vice versa.

Since there are both inefficiencies and inequities in education in many developing countries, it is possible to envisage situations where resources could be allocated so as to increase both efficiency and equity. In practice, however, it is not easy to identify such situations. Economists use the concept of a social welfare function to analyze the objectives of government policy—for example, efficiency, equity, and employment—but governments do not make explicit their welfare functions (in other words, the relative weights they give to different objectives.)

The tradeoffs that may have to be made between efficiency and equity have been studied in Brazil, where the geographical areas that need more resources on grounds of equity are not necessarily those in which social returns are highest (Behrman and Birdsall 1983a). In particular, poor rural areas have fewer and lower-quality schools than urban areas, but estimates of social rates of return in Brazil suggest that returns are higher in urban areas. If the government of Brazil was interested only in efficiency, it would invest in urban schools. If it was interested only in equity, it would invest in rural schools. Since it is trying to satisfy both efficiency and equity goals, how should resources be allocated? The answer depends to some extent on the relative priorities attached to efficiency and equity goals. The statements of politicians seldom throw much light on this question. Behrman and Birdsall have tried to identify the implicit objectives of the Brazilian government by looking at relative priorities in past expenditure patterns. The evidence here suggests that the Brazilian government has indeed tried to satisfy both objectives, but that somewhat more weight has been put on efficiency than on equity. If the government had shown no concern for equity, however, the geographical distribution of resources would have been even more unequal than it is at present.

This type of analysis demonstrates that efficiency-equity tradeoffs may exist in investment choices, but it cannot show which objectives should take precedence because such choices revolve around political rather than economic issues. The whole point of using social welfare functions is that they show the policymaker when it is possible to enhance both efficiency and equity, and when choices have to be made between investing in projects that offer the highest rates of return and those that would have the greatest redistributive impact. A recent study on financing education in developing countries emphasizes the political importance of the possible conflicts between equity and efficiency goals. Reviewing the evidence on education and income distribution in Colombia and Brazil, the study concludes that investing in primary education and financing that investment through progressive income taxation would have the greatest impact on equalizing benefits in the present generation, but suggests that this may have undesirable consequences for growth: "Thus there may be a trade-off between distribution and economic growth resulting from educational investment" (Carnoy et al. 1982, p. 65).

The Assessment of Equity Implications
of World Bank Projects

When the World Bank assesses educational investment projects, judgments are made about both efficiency and equity, but the question of tradeoffs is seldom explicitly considered. It is hoped that in the majority of cases investment will serve both efficiency and equity goals. The criteria for project assessments include indicators of internal efficiency, external efficiency (as judged by manpower and employment needs), and equity (as judged by social and geographical distribution of access).

There have been some significant improvements in equity as a result of Bank-financed projects. Recent projects in the Dominican Republic, Tanzania, and Tunisia, for example, have helped to achieve a broader social and geographical distribution of educational and training opportunities by increasing enrollment among the poorest groups in the population. In Chile, opportunities in rural areas have been extended as a result of mobile training units. Similar examples can be found among projects undertaken in Jordan, Cameroon, Kenya, Nigeria, and Malaysia. In addition, the Mobile Farmer Training Units in Kenya have provided access to training in areas with scattered populations. The degree of social equity achieved as a result of this wider distribution of facilities has not been established, however.

These are only representative examples of World Bank projects that have attempted to improve the distribution of educational opportunities. There have, however, also been some disappointments. Because of sociological factors, for example, male-female disparities have not been reduced as much as envisaged when projects were selected. In several cases rates of attrition have proved highest among the poorer students, and in some cases lack of scholarships has prevented low-income students from increasing their enrollment.

If educational investment is to contribute effectively to achieving equity as well as efficiency goals, the choice of project must be based on a careful assessment of the reasons underlying inequalities of access or opportunities. Such analysis may reveal that low access or high rates of wastage among certain groups are due to socioeconomic or cultural factors beyond the control of project planners. The political power of middle- and upper-class groups and elites and their determination to retain economic and educational privileges are often strong motivating forces in the provision of education. As Carnoy et al. (1982, p. 66) have noted, many of the suggestions for improving the equity of educational investment involve changes vehemently opposed by the middle class: "They are all good suggestions, but they have their political ramifications. Each solution . . . is a political solution, whose financial implications infer political conflict."

In evaluating educational investment, we must therefore take into account political and social realities as well as economic and financial criteria. We must also recognize that equity and efficiency tradeoffs have to be resolved within existing social and political as well as financial and practical constraints.

References

Adelman, Irma, and Cynthia Morris. 1973. *Economic Growth and Social Equity in Developing Countries*. Stanford, Calif.: Stanford University Press.

Ahluwalia, Montek. 1976. Inequality, Poverty and Development. *Journal of Development Economics* 3.

Alexander, Kern. 1982. Concepts of Equity. In *Financing Education: Overcoming Inefficiency and Inequity*, ed. Walter McMahon and Terry Geske. Champaign, Ill.: University of Illinois Press.

Altimir, Oscar. 1982. *The Extent of Poverty in Latin America*. World Bank Staff Working Paper no. 522. Washington, D.C.

Altimir, Oscar, and Juan Sourouille. 1980. Measuring Levels of Living Standards in Latin America: An Overview of Main Problems. Living Standards Measurement Study Working Paper no. 3. Washington, D.C.: World Bank, Development Research Department.

Anderson, C. Arnold. 1983. Social Selection in Education and Economic Development. Washington, D.C.: World Bank, Education Department.

Armitage, J., and R. H. Sabot. 1982. Educational Policy and Intergenerational Mobility: Analysis of a Natural Experiment. Washington, D.C.: World Bank, Development Research Department.

———. Forthcoming. Efficiency and Equity Implications of Subsidies of Secondary Education in Kenya. In *Modern Tax Theory for Developing Countries*, ed. David Newbery and Nicholas Stern. New York: Oxford University Press.

Atkinson, A. B. 1970. On the Measurement of Inequality. *Journal of Economic Theory* (August).

Becker, G. S., and B. Chiswick. 1966. Education and the Distribution of Earnings. *American Economic Review* (May):358–70.

Behrman, Jere, and Nancy Birdsall. 1983a. The Implicit Equity-Productivity Tradeoff in the Distribution of Public School Resources in Brazil. World Bank Country Policy Discussion Paper no. 1983-1. Washington, D.C.

———. 1983b. The Quality of Schooling: The Standard Focus on Quantity Alone Is Misleading. *American Economic Review* (December).

Birdsall, Nancy. 1982. The Impact of School Availability and Quality on Children's Schooling in Brazil. Population and Human Resources Division Discussion Paper no. 82-8. Washington, D.C.: World Bank.

Blaug, Mark. 1978. Thoughts on the Distribution of Schooling and the Distribution of Earnings in Developing Countries. Paper prepared for a seminar at the International Institute for Educational Planning, Paris.

———. 1982. The Distributional Effects of Higher Education Subsidies. *Economics of Education Review* (Summer).

Blaug, Mark, Christopher Dougherty, and George Psacharopoulos. 1982. The Distribution of Schooling and the Distribution of Earnings: Raising the School Leaving Age in 1972. *The Manchester School* (March):24–39.

Bowles, S. 1972. Schooling and Inequality from Generation to Generation. In *Investment in Education: The Equity-Efficiency Quandary*, ed. T. W. Schultz. *Journal of Political Economy* 80, no. 3:S219–51. Supplement.

Bowman, Mary-Jean, Benoit Millot, and Ernesto Schiefelbein. 1984. The Political Economy of Public Support for Higher Education: Studies in Chile, France, and Malaysia. Washington, D.C.: World Bank, Education Department.

Carnoy, Martin, in collaboration with Jose Lobo, Alejandro Toledo, and Jacques Velloso. 1979. *Can Education Policy Equalise Income Distribution in Latin America?* Farnborough, U.K.: Saxon House.

Carnoy, M., H. Levin, R. Nugent, S. Sumra, C. Torres, and J. Unsicker. 1982. The Political Economy of Financing Education in Developing Countries. In *Financing Educational Development*. Ottawa: International Development Research Centre.

Chenery, Hollis, and Moises Syrquin. 1975. *Patterns of Development 1950–1970*. Oxford: Oxford University Press.

Chiswick, Barry, and Jacob Mincer. 1972. Time Series Changes in Personal Income Inequality in the United States from 1939, with Projections to 1985. In *Investment in Education: The Equity-Efficiency Quandary*, ed. T. W. Schultz. *Journal of Political Economy* 80, no. 3:S34–66. Supplement.

Fields, Gary. 1975. Higher Education and Income Distribution in a Less Developed Country. *Oxford Economic Papers* (July).

———. 1978. *Assessing Educational Progress and Commitment*. Report for the U.S. Agency for International Development. Ithaca, N.Y.: Cornell University.

———. 1980. Education and Income Distribution in Developing Countries: A Review of the Literature. In *Education and Income*, ed. T. King. World Bank Staff Working Paper no. 402. Washington, D.C.

Fishlow, Albert. 1973. Distribuição de Renda do Brasil: Un Novo Exame. *Dados* no. 11.

Foxley, A., E. Aninat, and J. Arellano. 1976. Who Benefits from Government Expenditures? ILO Working Paper WEP 2-23 WP44. Geneva: International Labour Office.

———. 1977. The Incidence of Taxation. ILO Working Paper WEP 2-23/WP51. Geneva: International Labour Office.

Jallade, Jean-Pierre. 1973. *The Financing of Education: An Examination of Basic Issues*. World Bank Staff Working Paper no. 157. Washington, D.C.

———. 1974. *Public Expenditures on Education and Income Distribution in Colombia*. Baltimore, Md.: Johns Hopkins University Press.

Jencks, Christopher. 1972. *Inequality*. New York: Basic Books.

Knight, J. B., and R. H. Sabot. 1982. Educational Expansion, Wage Compression and Substitution between Educational Levels: A Quantitative and Policy Analysis. Washington, D.C.: World Bank, Development Research Department.

Kuznets, Simon. 1955. Economic Growth and Income Inequality. *American Economic Review* (March).

———. 1956. Quantitative Aspects of the Economic Growth of Nations. *Economic Development and Cultural Change* 5.

Langoni, Carlos G. 1972. Distribuição de Renda e Desenvolvimento Economico do Brasil. *Estudios Economicos* (October).

Maas, Jacob, and Jacob Criel. 1982. *Primary School Participation and Its Internal Distribution in Eastern Africa*. World Bank Staff Working Paper no. 511. Washington, D.C.

Marin, A., and G. Psacharopoulos. 1976. Schooling and Income Distribution. *Review of Economics and Statistics* 58, no. 3 (August):332–38.

McMahon, Walter. 1982. Efficiency and Equity Criteria for Educational Budgeting and Finance. In *Financing Education: Overcoming Inefficiency and Inequity*, ed. Walter McMahon and Terry Geske. Champaign, Ill.: University of Illinois Press.

Mingat, A., and J. P. Tan. 1985. On Equity in Education Again: An International Comparison. *Journal of Human Resources* 20 (Spring):298–308.

———. Forthcoming. Who Profits from the Public Funding of Education? A Comparison by World Regions. *Comparative Education Review*.

Psacharopoulos, George. 1977. The Perverse Effects of Public Subsidisation of Education. *Comparative Education Review* 21, no. 1 (February):69–90.

———. 1978. Inequalities in Education and Employment. A Review of Key Issues with Emphasis on LDC's. International Institute for Educational Planning Working Document IIEP/549/8A. Paris.

Psacharopoulos, George, and William Loxley. Forthcoming. *Diversified Secondary Education and Development: Evidence from Colombia and Tanzania*. Baltimore, Md.: Johns Hopkins University Press.

Rogers, Daniel. 1971. Financing Higher Education in Less Developed Countries. *Comparative Education Review* 15:20–27.

Schultz, T. W., ed. 1972. *Investment in Education: The Equity-Efficiency Quandary*. Supplement. *Journal of Political Economy* 80, no. 3.

Selowsky, Marcelo. 1979. *Who Benefits from Government Expenditure? A Case Study of Colombia*. New York: Oxford University Press.

Sen, A. K. 1973. *On Economic Inequality*. New York: Oxford University Press.

ter Weele, Alexander, Jacob Maas, Jose Dominguez-Urosa, and John Villaume.

1979. Equity and the Distribution of Education Services: Representation Index and Other Techniques for Policy Analysis and Project Identification. Economic Development Institute Case Study and Exercise Series. Washington, D.C.: World Bank.

Thurow, L. C. 1975. *Generating Inequality*. New York: Basic Books.

Winegarden, C. R. 1979. Schooling and Income Distribution: Evidence from International Data. *Economica* (February).

World Bank. 1980a. *Education*. Sector Policy Paper. Washington, D.C.

———. 1980b. *World Development Report 1980*. New York: Oxford University Press.

———. 1981. *World Development Report 1981*. New York: Oxford University Press.

10

Intersectoral Links

The evaluation of education investments becomes even more compli-
cated when we consider that yet another factor must be taken into
account—the important links between education and other sectors.
Education may be linked, for example, with agricultural development,
health, and other forms of human resource development.

The World Bank recognizes the importance of intersectoral links for
development, as will become clear in this chapter. The discussion begins
with a brief review of the Bank's approach to the problems of alleviating
poverty and to the relationships between investments in education,
health, nutrition, and the reduction of fertility. The link between education and health and fertility is examined in some detail. In addition, we
look at the importance of education and training components in investment projects in other sectors, including agriculture, transport, mining,
and irrigation.

An Intersectoral Approach
to the Alleviation of Poverty

Many World Bank reports and publications have explained why the
attack on poverty became a fundamental goal of development policy
during the 1970s. The 1980 *World Development Report*, for example,
noted that the underlying purpose of the attack on poverty itself was to
increase employment, meet basic needs, reduce inequalities in income
and wealth, and raise the productivity of the poor. The report argued
convincingly that human resource development should be the main instrument used to achieve these goals (World Bank 1980, p. 46).

Human development encompasses education and training, better
health and nutrition, and the reduction of fertility, each of which makes
its own important contribution to human development. Together these
elements form a "seamless web" (see figure 10-1) in which the links and
relationships considerably enhance the productivity of investment in
education. That is to say, improvements in education can help alleviate

Figure 10-1. *Policy and Poverty*

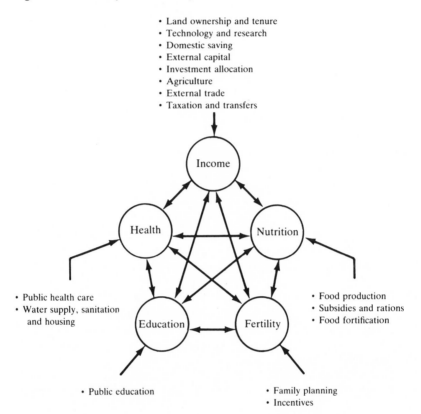

• Land ownership and tenure
• Technology and research
• Domestic saving
• External capital
• Investment allocation
• Agriculture
• External trade
• Taxation and transfers

Income

Health   Nutrition

• Public health care   • Food production
• Water supply, sanitation   • Subsidies and rations
  and housing   Education   Fertility   • Food fortification

• Public education   • Family planning
                                       • Incentives

Source: World Bank (1980), p. 69.

poverty both directly and indirectly by increasing income, improving health and nutrition, and reducing family size. Healthy children learn more effectively than sick children, well-nourished children learn more effectively than hungry children, and educated parents are more likely to have healthy and well-nourished children. Thus the two-way relationships depicted in figure 10-1 represent the benefits to be gained from investment in education. These include better health and nutrition and lower fertility, which in turn will make future educational investments more productive. Recognition of these crucial relationships was one of the main justifications for the Bank's increased emphasis during the 1970s, in both lending and research, on meeting basic needs as an approach to economic development (ul Haq and Burki 1980; Streeten 1981; Isenman et al. 1982).

Many examples of the relationships between education and other forms of social investment can be cited (Noor 1981). Clean water can make an important contribution to better health, for example, but whether it will do so depends on the relevant education and understanding of its users (Feachem et al. 1983). Before education can make investment in improved health or sanitation more productive, however, there must be appropriate synchronization between investment in education and other cross-sectoral efforts. In general, investment in education should parallel or follow other sectoral investments. New knowledge about the effects of contaminated water, for example, will be of little use if no alternative supply is available; conversely, new wells may be of little use if the community knows nothing about sanitation and hygiene (Noor 1981, p. 43).

A number of studies on basic needs (for example, Streeten 1981) illustrate how intersectoral links may provide a powerful justification for investing in basic education. The evidence on the contribution of primary schooling to economic development (Colclough 1980) indicates that the interactive or strengthening effects of schooling on various aspects of social policy—including family size, health, nutrition, literacy, and awareness of national culture—reinforce the economic case for investing in primary education. Furthermore, it has been suggested that an analysis of basic needs may provide investments in basic education with a host of new justifications from other fields, such as demography, health, nutrition, and rural development, where progress can be achieved only when basic educational needs are met (Jallade 1982, p. 5).

As noted in chapter 3, comparisons of the costs and benefits of educational investments must include measures of the indirect benefits of education with respect to health, nutrition, or fertility, as well as the direct effects on earnings. The links between education, health, and fertility provide one way of measuring the nonmarket effects of education, which, according to recent estimates in the United States, amount to as much as two-fifths of the full economic value of education (Haveman and Wolfe 1984). It could be argued that in developing countries the ratio of nonmarket to market effects is even greater than in developed countries.

Furthermore, analysis of the effectiveness or output of education (see chapter 8) must include measures of the effects of education on other sectors. For example, cost-effectiveness studies of different curricula or examination methods or of the effects of inputs such as better-trained teachers or new textbooks could take into account the contribution of education to improved health or nutrition. Kenya provides an excellent example of how attempts to improve the internal and external efficiency of education may strengthen the links between education and other sectors and thus enhance the social benefits of educational investment. By reforming its primary-school curriculum and the certificate of primary education (CPE), Kenya has hoped to increase the relevance of primary-school teaching and make a greater contribution to knowledge about health and nutrition.

In view of the importance of intersectoral links, any attempt to evaluate the impact of investment undertaken to satisfy basic needs should be concerned with the overall impact on poverty as well as with the other effects of individual projects. So far, few such evaluations have been made, although Jallade (1982) has proposed that the impact of the Bank's education lending be evaluated to assess both its overall effectiveness and the direct and indirect effects on the poor. Since education allows people to make the most of their investments in other sectors, such as family planning, rural development, or health, a wide variety of long-term and short-term effects must be measured, but evidence on such effects is usually lacking. It might be possible to obtain some of the necessary data through a poverty impact evaluation study, which would encompass all projects already implemented in a single country and look at their interrelationships as well as their direct effects. In the absence of such a study, however, some of the relationships between education and other sectors can still be analyzed.

Education and Health

The relationships between education and health have been extensively reviewed by Cochrane, O'Hara, and Leslie (1980), whose principal concern has been the links between education and life expectancy and between education and child mortality, disease, and nutrition. Cochrane's analysis begins with the fact that the rate of reduction of mortality in developing countries seems to have slowed, and that levels of life expectancy are still below what was thought achievable only a few years ago (Gwatkins 1979). As a result, Cochrane points out, considerable research has looked into the determinants of mortality. This work has shown that the level of mortality is the product of two factors: the level of knowledge regarding ways to combat diseases, and the means available for implementing that knowledge. It has also shown that education helps to determine both the level of knowledge and the ease with which it can be transmitted and utilized. According to cross-national studies of the factors correlated with life expectancy, literacy and other measures of education are more closely related to life expectancy than per capita income or even the number of doctors per capita. There is evidence that education affects health, and therefore mortality, both directly and indirectly (Cochrane, O'Hara, and Leslie 1980). One of the indirect effects is that one person's education may influence the health of another.

An important link has been found between parental education, particularly the level of a mother's education, and a child's health. Among the numerous studies of the impact of parental education on child mortality and on child nutrition, those in Latin America, Africa, and the Middle East all show a significant positive association between a mother's educational level (or literacy) and her child's weight or height, which are used as a measure of nutrition. Similarly, a strong relationship exists between a mother's educational level and her child's life expectancy. Studies in Latin America, Asia, Africa, and the Middle East, which correlate parental education with the probability of a child surviving after the age of two, show that the children of educated mothers have a far higher survival rate. The evidence is unequivocal: educated parents, particularly mothers, have better-nourished children who are less likely to die in infancy than the children of uneducated parents. On the average, one additional year of schooling for a mother results in a reduction of 9 per 1,000 in child or infant mortality.

The way in which education leads to improved health is not yet fully understood, although it seems clear that education is associated with well-nourished children in part because better-educated parents have higher incomes. Education affects nutrition even when income is held constant, however; indeed, some studies suggest that income differences explain less than half the observed differences in nutrition of children, and that education is the most important factor. The effects of education may not be entirely positive. In some cases, educated mothers are less likely to breast-feed, for example, because they are more likely to be employed. The relationship between education and nutrition and health obviously needs further analysis before it will be possible to say how education can best contribute to improvements in child health.

Information about health and nutrition can be disseminated in many ways, of course, and need not be linked with formal education, as can be seen in some of the primary health care systems involving community health workers that have been an important trend during the 1970s. Following the example of China's "barefoot doctors," countries as diverse as Iran, Brazil, Sudan, India, Jamaica, Botswana, and Tanzania have introduced primary health care systems of this type. The training of these workers is crucial, of course, but experience in the Sudan shows that the most effective policy is not necessarily to recruit people with formal qualifications, who tend to be young and are not always easily accepted by local communities. Recognizing the value of maturity and personal exerience, many programs now prefer to recruit older, highly motivated people, rather than younger, better-educated applicants (World Bank 1980, pp. 58, 74).

Table 10-1. *Tanzania Primary-School Leaving Examination:*
Selected Science Items, 1978

---

36. A frog can live both on land and in water because
    A. it has eyes that enable it to see while in water and on land
    B. it has lungs and skin which it uses for breathing while on land and in water
    C. it has special legs that enable it to swim in water
    D. its food is found on land and in water
37. Plant roots grow towards
    A. light
    B. gravitational pull
    C. water
    D. soil
38. Iron is prevented from rusting by painting it with
    A. petrol
    B. kerosine
    C. grease
    D. oil
39. The requirement for eggs to hatch is usually
    A. constant heat
    B. constant light
    C. presence of mother to warm them
    D. a machine made for the purpose of hatching eggs
40. An instrument that makes use of properties of light is
    A. a looking glass
    B. a torch
    C. a camera
    D. sun goggles

Source: Somerset (1982).

Nevertheless, the formal education system can contribute to primary health care in many ways. As mentioned earlier, a case in point is Kenya, which reformed its examination system to increase the relevance of the knowledge tested in the certificate of primary education (CPE) examination (Somerset 1982). Another goal of this reform was to encourage primary school teachers to change both their teaching methods and the content of their lessons so that pupils, particularly in the final year of primary school, might obtain knowledge and skills of direct relevance to their future lives as citizens. The main purpose of many examinations at the end of the primary cycle is to select pupils for secondary schools, and as a result these examinations sometimes pay virtually no attention to testing knowledge (for example, about health and nutrition) that will be useful in the future lives of those who leave school after the primary level. This situation has an unfortunate backwash effect on the amount of time teachers are prepared to devote to these important topics during the final years of primary schooling. Yet the ability of education to contribute to better health or living standards could be significantly enhanced if examinations, and therefore teachers, attached higher priority to instruction in those areas.

It is interesting to compare the questions that are typically set in science papers in Tanzania, where there has been no examination reform (see table 10-1) with those recently set in CPE examinations in Kenya (table 10-2). The Tanzanian questions focus on physical science and formal biology, rather than on agriculture, nutrition, or health. In contrast, questions in the CPE examination in Kenya are intended to test the knowledge directly relevant to the attempt to improve standards of health and nutrition.

Many questions in Kenya's examination are specifically designed to help teachers improve pupils' knowledge or awareness of health and nutrition issues. For example, in recent years it has become increasingly common for mothers in low-income families in Kenya to favor bottle-feeding. The team responsible for setting the questions in the science CPE therefore considered it important that school leavers should know the main advantages of breast-feeding. Thus one of the questions in the 1981 CPE examination was as follows:

A standard VII class suggested the following reasons why mothers should feed their babies with breast milk rather than with bottled milk:

i) Breast milk is easier for young babies to digest.
ii) Babies who drink bottled milk are more likely to suffer from diarrhoea.
iii) Bottled milk does not contain enough water for young babies.
iv) It is cheaper to feed a baby with breast milk.
   Which of these are CORRECT:
a. (i) and (ii) only
b. (iii) and (iv) only
c. (i), (ii) and (iv) only
d. (i), (ii) and (iii) only

The format of the questions is also important. Of the four statements about feeding, only one is incorrect, and the candidate must identify this statement. This mode of presentation is used frequently with knowledge questions in the new CPE papers in Kenya in order to enhance their usefulness as a teaching resource. Such an approach is particularly important in developing countries where up-to-date textbooks and other learning resources are in short supply in the schools and teachers have to use obsolete materials or none at all. The importance of developing links between education and primary health care and nutrition is already being recognized in developing countries, and should be taken into account in evaluating and, in some cases, designing education projects.

Table 10-2. Kenya Primary-School Leaving Examination: A Selection of Science Items, 1975–1979

1975 Good farmers dip or spray their cattle to kill cattle ticks. The main reason they do this is because

    A. ticks make holes in the skin of cattle and this reduces the value of the leather

    B. ticks feed on the blood of cattle, which makes them weak

    C. cattle ticks carry diseases which are dangerous to human beings

    D. ticks carry diseases which can kill cattle

1977 There is a swamp at the far end of Muturi school compound. Kithui's home is on the other side of the swamp. If he walks through this swamp on his way to school each day, which one of the following diseases is he most likely to get?

    A. Bilharzia

    B. Kwashiorkor

    C. Rickets

    D. Smallpox

1977 Koske's mother gave four reasons why she wanted to replace the old grass roof on her house with a new *mabati* (corrugated iron) roof. Three of her reasons are correct. Which one is WRONG?

    A. A *mabati* roof keeps the house cooler

    B. Snakes and rats cannot hide in *mabati* roofs easily

    C. Clean water can be collected from the *mabati* roof

    D. *Mabati* roofs last longer than grass roofs

1977 Less than 3% of Kenya's land area is covered with forests. Four possible reasons why parts of Kenya should be kept for forests are given below. Only three of the reasons are correct. Pick out the one which is NOT correct.

    A. Forests help prevent soil erosion

    B. Forests help the soil to store water

    C. Forests provide raw materials for paper manufacture

    D. Forests help prevent the spread of sleeping sickness

1978 There was an outbreak of foot-and-mouth disease on Katuko's farm. She was told not to move her cattle from her farm. The main reason for this was that

    A. it would be easier to sell cattle on the farm

    B. sick cattle would die if moved

    C. the cattle would become healthy if they remained on the farm

    D. moving the cattle off the farm would spread the disease

1979 Cut dry grass is often placed between rows of coffee trees. Three of the following are reasons for doing this. Which one is NOT a reason?

    A. The grass reduces the speed with which the rain strikes the soil.

    B. Water evaporates from the soil more slowly

    C. The grass protects the coffee beans from pests.

    D. The grass helps to prevent weeds from growing around the coffee trees.

1979 Diarrhoea is a disease that kills many babies in Kenya. When babies have diarrhoea they lose a lot of water and foods. One correct way to treat babies with diarrhoea is to

    A. keep them wrapped up and warm so that they sweat out the sickness

    B. give them drinks of boiled cold water containing some sugar and a little salt

    C. give them solid foods containing plenty of carbohydrates

    D. give them very little food or water until the diarrhoea stops

1979 Wangari lives in a village with many people. From which one of the following sources can Wangari collect the best drinking water?

    A. A big river nearby.

    B. The rain from her mabati roof.

    C. The dam near the village.

    D. A swamp on her farm.

Source: Somerset (1982).

Education and Fertility

Another important link that has long been recognized but only recently examined in detail is the one between education and fertility. One of the indirect benefits of education in developing countries is often a reduction in fertility (see chapter 3). The relationship between education and fertility is a complex one, however, and in some circumstances the expansion of education in a developing country may actually result in increased fertility. Although many studies have demonstrated close links between education and fertility, the mechanisms by which education affects fertility have not received much attention. They were therefore the main focus of a recent review (Cochrane 1979) summarizing the ample evidence that education is related to fertility.

The study explored how education affects fertility and how it influences the direct determinants of fertility, such as age of marriage, desired family size, and knowledge of contraceptives. The study showed that the relation between education and fertility is far from uniform. In some countries education appears to be either unrelated to fertility or actually positively related. This means that before the relationship between education and fertility can be used for policy purposes, the question of when and why an inverse relation is likely to arise must be resolved.

The many ways in which education affects individuals are summarized in figure 10-2. The externalities, or spillover effects, of education in the

Figure 10-2. *Multiple Effects of Individual Education*

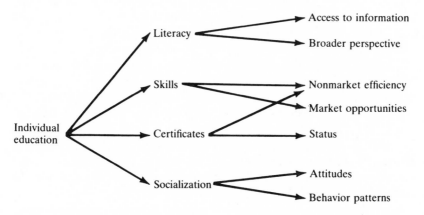

Source: Cochrane (1979), p. 29.

Figure 10-3. *Effects of Individual and Community Education*

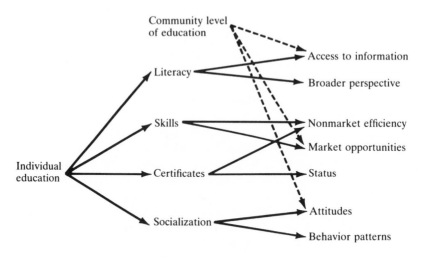

Source: Cochrane (1979), p. 30.

community as a whole are illustrated in figure 10-3. Research on education and fertility is now attempting to analyze the effects of individual and community-level education on individual fertility, but so far the results are limited. Cochrane's review of studies at the individual level, which provides evidence from more than twenty developing countries, suggests that literacy is by and large associated with reduced fertility, but that various patterns exist (see figure 10-4). The relationship between education and fertility is more likely to be inverse in urban than in rural areas, and there is also some evidence of thresholds. In countries with illiteracy rates greater than 60 percent, for example, individuals with some education appear to have higher fertility than those with no education, whereas in countries where illiteracy is less than 40 percent, such individuals tend to have lower fertility (Cochrane 1979). These data suggest an interaction between aggregate level of education and individual level of fertility. The fact that there may be such thresholds means that it is very difficult to predict the effects of education on fertility in countries with high levels of illiteracy. Since there may be tradeoffs between short- and long-term effects, the study concludes that appropriate policy cannot be determined unless more is known about the precise ways in which education influences fertility (Cochrane 1979).

Education affects fertility indirectly through a number of intervening variables (figure 10-5). These variables fall into three categories: factors

Figure 10-4. *Relation between Education and Fertility*

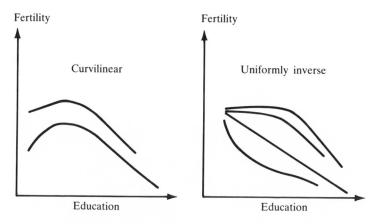

Source: Cochrane (1979), p. 39.

that determine the demand for children, factors that determine the supply of children, and fertility regulation factors.

According to evidence from more than fifty studies, education may affect many of these variables. It affects the demand for children, for example, by changing the perceived costs and benefits of having children and the ability to afford children, and also by altering preferences, as reflected in ideal family size. It also affects the supply of children by affecting the age of marriage, the health of both parents and children, and the likelihood that newborn infants will survive. Finally, education affects the regulation of fertility by increasing communication between husband and wife, by increasing knowledge of contraceptives, and possibly, though not always, by changing attitudes toward contraception. Studies in Latin America suggest that education affects contraceptive use through its effects on knowledge about contraception, attitudes, motivation, and access to family planning services.

As can be seen from table 10-3, some of the effects are positive, whereas others are negative. Furthermore, education of males may have a different effect from education of females. In general, the positive effects are stronger in the case of a husband's education, and the negative effects are usually stronger in the case of a wife's education. A number of general conclusions can be drawn from this evidence:

- Education cannot be expected to reduce fertility in all circumstances. In particular, in the poorest and least-literate societies, small amounts of education may actually lead to higher fertility initially.

Figure 10-5. *Model of the Effect of Education on Fertility through the Major Intervening Variables*

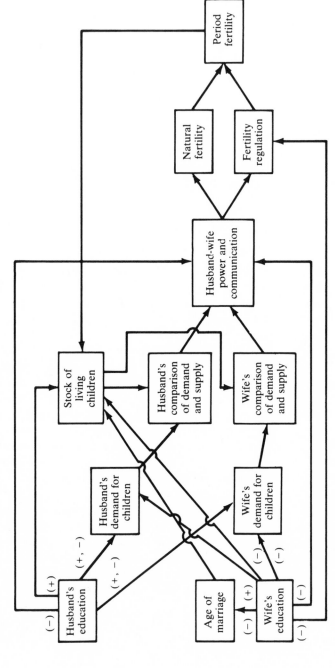

Source: Cochrane (1979), p. 145.

Table 10-3. *Evidence Supporting the Relation between Education and Fertility through the Intervening Variables*

Variable	Relation of education and variable	Probable relation of education through the variable	Results Supporting (number of cases)	Results Not supporting (number of cases)
Potential supply of births				
Probability of marrying	inverse	−	6	5[a]
Age of marriage	direct	−	59	12[a]
Health	direct	+	2	0
Lactation	inverse	+	6	0
Postpartum abstinence	inverse	+	2	0
Infant or child mortality	inverse	−	16	7
Demand for children (desired family size)	inverse	−	17	8[a]
Preference for children				
Ideal family size	inverse	−	20	7[a]
Desired number of sons	inverse	−	8	1
Perceived benefits of children	inverse	−	17	2
Perceived costs of children	direct	−	2	0
Perceived ability to afford children	direct	+	9	3
Fertility regulation (contraceptive use)	direct	−	26	11
Attitudes toward birth control	direct	−	28	4
Knowledge of birth control	direct	−	28	1
Husband-wife communication	direct	−	9	0

a. Relation of male education to the variable is much weaker than that of female education.

Source: Cochrane (1979), p. 146.

- There is tentative evidence that over time education will ultimately reduce fertility.
- Increasing female education will be more likely to reduce fertility than increasing male education.
- Education is more likely to reduce fertility in urban than in rural areas.

These conclusions are important for evaluations of education investment, but they do not point to a single educational strategy that would maximize the effectiveness of education in reducing fertility. Some evidence suggests that female education should receive priority, but it is unclear whether a broad program of elementary education should be recommended or a narrower program of secondary education. Once again, there are tradeoffs. In the least-literate countries, secondary education is more immediately effective in reducing fertility but has a higher cost and therefore implies a more limited distribution of educational benefits than would be the case if primary education were expanded. Although mass literacy campaigns may increase fertility in the short run by improving standards of health, they will no doubt reduce fertility in the long run by changing attitudes toward birth control.

Much of the evidence has to do with the effect of parental education on fertility, but it is also possible that the schooling of children is linked with fertility. Among the influences on fertility shown in figure 10-5 are the husband's and wife's demand for children, which may be governed by the costs associated with the upkeep of children, including the costs (direct and indirect) of education. Direct costs include expenditure on fees, books, and other such items, while indirect costs include earnings forgone.

It has been suggested (Caldwell 1982) that fertility will decline when parents begin to perceive that the costs arising from an additional child, owing to the direct and indirect costs of education, outweigh the financial benefits expected from the child, for example, in the form of security for the parents in old age. Caldwell's study concentrates on the effect of mass education on family attitudes and perceptions, that is, on the "family economy." Investment in mass education may change attitudes and perceptions and thus reduce fertility in at least five ways:

- The child's potential for work inside and outside the family may be reduced (that is, indirect costs to the family may be increased).
- Direct costs of education (for example, fees, books, uniforms, and so on) may be increased.
- Rather than being regarded as a present productive resource within the family, the child may be regarded as a future productive resource by society and the family; this change in status is often accompanied

by laws to protect children, such as child labor laws and compulsory schooling legislation.

- Mass education may accelerate cultural change and the creation of new cultural values.
- Schools serve as a major instrument for propagating new ideas and values.

The first two mechanisms have been widely cited in the literature and there is considerable evidence of the influence of both direct and indirect costs of education on fertility. Caldwell argues, however, that the last three value-related mechanisms are more influential in turning the family economy from one in which children are perceived as a net benefit (in financial terms) to one in which they are seen as a net cost. The importance of children's education on parental fertility is also evident from the data in table 10-4, which show that among 37 developing countries, the primary enrollment rate in all was at least 80 percent at the date of the historical decline in fertility. The secondary gross enrollment was also high, averaging 33 percent (Bulatao 1983).

Tan and Haines (1984) examine the historical experience of developed countries, particularly that of the United States and Europe in the nineteenth century, along with the recent experience of developing countries, in order to assess the influence on fertility of changes in general enrollment rates. The evidence suggests that in developed countries a substantial level of primary-school enrollments (50–60 percent of the school-age population) has to be attained before fertility declines significantly. A much lower secondary-school enrollment rate is associated with fertility decline, but considerable variations in secondary-school enrollment rates were observed in developed countries at the time fertility began to decline sharply.

In the case of developing countries, a comparison of gross enrollment rates (GER) and fertility rates suggested the following trends:

- With few exceptions, the primary-school GER must be at least 75 percent before a sustained decline in fertility can be discerned.
- The decline in fertility is most significant in countries in which the primary GER has already stabilized beyond the 100 percent level, as has been the case in Chile, Costa Rica, Barbados, Singapore, Korea, and Hong Kong.
- As the primary GER increases, fertility may also increase, as was the case between 1950 and 1960 in Costa Rica, Chile, and Hong Kong.
- In quite a few countries, increases in the primary GER were not accompanied by reductions in the total fertility rate, particularly when enrollment was rising from low levels. This relationship is apparent for various periods in Colombia, Mexico, India, Egypt, Tunisia, and most notably in Kenya and Ghana.

Table 10-4. *Gross Enrollment Levels at the Start of Fertility Decline*
in Selected Developing Economies

Economy	Start of fertility decline[a]	Gross enrollment rate at start of fertility decline[b]	
		Primary	Secondary
American Samoa	1964	92	85
Bahamas	1970	153	76
Bahrain	1966	118	42
Barbados	1963	80	36
Brazil	1962	108	15
Kampuchea, Dem. (Cambodia)	1968	84	10
Cape Verde	1961	70	8
Chile	1961	105	25
Colombia	1962	83	14
Costa Rica	1961	100	27
Cuba	1967	118	25
Dominican Republic	1966	91	15
Fiji	1966	85	20
Grenada	1964	156	21
Guadeloupe	1968	148	37
Guam	1961	100	90
Guyana	1961	117	44
Hong Kong	1961	92	30
Jamaica	1971	85	59
Korea	1959	96	27
Lebanon	1965	104	22
Malaysia	1962	84	27
Martinique	1969	146	47
Mauritius	1960	98	24
Mexico	1971	104	22
Panama	1968	103	39
Philippines	1969	114	50
Puerto Rico	1961	114	70
Reunion	1960	126	20
Singapore	1955	94	18
Sri Lanka	1963	94	31
Suriname	1972	100	38
Thailand	1968	81	13
Trinidad and Tobago	1960	128	14
Tunisia	1968	107	20
Turkey	1972	106	29
Venezuela	1963	95	30
Average	—	105	33

a. *Source:* Bulatao (1983).
b. *Source:* Unesco (1970, 1977).

- A downward trend in the total fertility rate is sometimes unaccompanied by increases in primary-school enrollment, as in Thailand and Egypt.
- A negative relationship between schooling and fertility trends is more frequently observed when the schooling indicator is secondary rather than primary enrollment, as was the case in Singapore, Korea, and Chile.

Another approach to explaining the link between parental fertility and children's education has been taken by economists. The earliest formulation of this link is Becker's quantity-quality tradeoff model (Becker 1960). Child schooling is considered an investment in human capital that enhances child quality. In the simplest form of the model, the demand for both the number (quantity) and quality of children depends on the shadow prices of quantity, quality, and other commodities, and on shadow income. An increase in the demand for child quality reduces the demand for quantity, both through the shadow income constraint and by raising the shadow price of quantity. An increase in the demand for quantity reduces the demand for quality by a similar mechanism. Becker (1981, pp. 106–07) suggests that this interaction between quantity and quality offers the "most promising explanation" for the "large decline in births in the United States during the twenties or in Japan during the fifties."

The evidence from most of the studies of the links between enrollment and declining fertility in developing countries suggests that a sharp fall in fertility appears to coincide with rising secondary-school enrollment in an environment where universal primary education was either already established or close to achievement. The concept of a quantity-quality tradeoff is corroborated by data from sixty-nine developing countries in the period 1968–70, which indicate that a 10 percent rise in the secondary-school enrollment ratio is accompanied by a 1.2 percent drop in fertility. At the early stages of development, however, fertility is much less sensitive to socioeconomic changes, including increases in school enrollment, than to influences such as improvements in life expectancy at birth (Anker 1978).

The analysis of the relationship between parental education and fertility and between school enrollment and fertility shows that the links between education and fertility are far more complex than were assumed in the past. Not only are the direct and indirect costs of education significant determinants of demand for children, but the way in which these costs are shared between parents, the extended family (including siblings), and government influences parents' perceptions of the economic burden (or advantage) of an additional child.

Thus, all the evidence of a strong, but complex set of relationships between parental education, children's education, and fertility and the relationships between education, health, and nutrition indicate that decisions about education projects need to consider both indirect and direct effects, even though these are difficult to predict precisely. Furthermore, efficiency-equity tradeoffs (see chapter 9) have bearing on the indirect as well as the direct effects of educational investment. The fact that female education appears to have a greater effect on reducing fertility than does male education, for example, gives strong support to the general arguments for reducing education disparities, whereas the fact that education appears to have more effect on fertility in urban than in rural areas may suggest the opposite conclusion. Education in rural areas has other important intersectoral links, however.

Education and Agricultural Development

Considerable evidence indicates that education raises the productivity of farmers (see chapter 3). Studies in Korea, Malaysia, Nepal, and Thailand demonstrate that education increases the physical productivity of farmers and that the effects of education on crop yields and other physical measures of farm output are "positive, statistically significant and quantitatively important" (Jamison and Lau 1982, p. 8). Four years of schooling, on the average, appears to increase the output of farmers by about 8 percent. Estimates of rates of return to rural education in Korea, Malaysia, and Thailand confirm that education is a profitable investment. A calculation of the returns to a World Bank education project in Thailand, for example, suggests a rate of return of at least 20 percent.

The economic justification for investing in education in rural areas is strengthened even further when the relationships between education and other ways of boosting agricultural productivity are taken into account. Schultz (1964) has suggested that education will be more effective in a changing, modernizing environment than in a traditional one. World Bank experience strongly supports this hypothesis, particularly when the effects of education on farmers' efficiency in traditional environments (characterized by primitive technology, traditional farming practices and crops, and little reported innovation or exposure to new methods), are compared with the effects in a modern environment (characterized by new crop varieties, innovative planting methods, erosion control, and modern inputs such as insecticides, fertilizers, and tractors or machines). That the effects of education are substantially greater in the modern environment can be seen from the mean increase in output after four years of education: it is 1.3 percent in traditional conditions, but 9.5 percent in modern or modernizing conditions (Jamison and Lau 1982, p. 38).

There is a two-way relationship between education and other investments to improve agricultural productivity. The effect of education on farmers' output is considerably greater when complementary inputs such as fertilizers or new types of seed are available; at the same time, education, particularly literacy and numeracy, is often a prerequisite for the introduction of modern farming methods that rely on fertilizers and new types of seed. There appear to be four basic stages of agricultural technology, and each has its minimum learning requirements (figure 10-6). Traditional farming, where techniques are handed from father to son, requires little or no formal education. The second stage involves the use of a single modern input; for example, the utilization of fertilizer is considerably improved if farmers have rudimentary literacy and a knowledge of addition, subtraction, and division. In the third stage, which uses several complementary inputs simultaneously, technology can be aided if farmers have more complex mathematical skills and a rudimentary knowledge of chemistry and biology. Finally, full irrigation-based farming requires farmers to calculate the effects of changes in crops, climate, and so on (Heyneman 1983).

Thus education appears to be a prerequisite of investment in irrigation projects, and seems to increase the yield from other types of agricultural investment. There is also evidence that formal education of farmers may improve the effectiveness of investment in agricultural extension services. This raises the question of whether formal and nonformal education, including agricultural extension services, are substitutes or complements.

The few studies that have attempted to measure the interaction between formal and nonformal education have found positive coefficients of interaction, which suggest a possible complementary relation between the two forms of education. There is no firm evidence on the extent of the interactions, however, nor on the precise way in which education increases agricultural productivity. Results from World Bank research in this area in Nepal and in Thailand should show whether interactions between the two types of investment provide a strong economic argument for increased investment in education, and which types of educational investment are likely to be most profitable. The evidence thus far points to a linkage between education, agricultural extension services, the use of modern agricultural techniques, and other aspects of rural development. The precise mechanisms by which these links are effected are still unknown, but one significant piece of evidence on the importance of the

Figure 10-6. *Four Basic Stages of Agricultural Productivity and Their Learning Requirements*

Farmer-entrepreneurs' technology level	Agricultural inputs	Minimum learning requirements
Level A: Traditional farming techniques passed from parent to child	Local varieties of seeds and implements	Addition and subtraction—not necessarily acquired through formal education
Level B: Intermediate technology	Small quantities of fertilizer	Addition, subtraction, division, and rudimentary literacy
Level C: Fully improved technology	High-yielding varieties; proven seeds; rate of application of seed, fertilizer, and pest control per acre	Multiplication, long division, and other more complex mathematical procedures; reading and writing abilities, and rudimentary knowledge of chemistry and biology
Level D: Full irrigation-based farming	All above inputs; tubewell access during the off-season; and water rates per acre	Mathematics, independent written communication, high reading comprehension, ability to research unfamiliar words and concepts; elementary chemistry, biology, physics; and regular access to information from print and electronic sources

Source: Heyneman (1983)

links between education and the productivity of agricultural investment projects comes from a recent study of the effects of project-related training in the agricultural sector.

The Education and Training Components of Investment in Other Sectors

Not all the investment in education that is financed through the World Bank takes place as a result of projects in the education sector. A substantial amount of education and training is a basic component of investment projects in other sectors, such as transport, power, or agricultural and rural development. In 1978 the World Bank estimated that such project-related training might amount to as much as a quarter of total investment in projects in the education sector. In order to assess the importance of such training and evaluate its profitability and contribution to the development process, the Bank's Operations Evaluation Department (OED) together with Unesco and ILO carried out a review of training in Bank-financed projects that looked at forty-two training components in twenty projects outside the education sector (World Bank 1982). More recently, a study of project-related training in fifty-two agricultural projects demonstrated that the amount of training provided in a project has a powerful influence on the economic success of the investment (Mingat 1984).

Most project-related training has two important characteristics: it generally accounts for a small percentage of total project costs, and the economic efficiency or output of the project is usually assumed to depend heavily on the skills of the people who work on the project. This suggests that the profitability of investment in training is high in relation to the total costs of a project. According to Mingat's (1984) analysis, the rate of return to project-related training is very high; the rate of return on training is likely to be at least 50 percent and possibly more than 100 percent (depending on the opportunity costs of the training).

Closer analysis of the data for individual projects demonstrates that training and literacy may be complementary in the sense that the effects of project-related training are greater in countries with a high general level of literacy. Moreover, training has little effect when the adult literacy rate is less than 40 percent. Mingat suggests that the poor economic performance of agricultural projects in sub-Saharan Africa is partly due to the low literacy rate in the region. Mingat also shows that the size of the training component is an important determinant of the ultimate success of agricultural investments and that it is more significant than GNP per capita, for example. The links between the effectiveness of training and the general literacy level in a country suggest that training is most effective in countries with literacy rates well above 40 percent.

Mingat thus provides proof of the complex links between investment in general education, specific training, and investment in other sectors, particularly agriculture. He also argues that project evaluations should spend more time looking for the most cost-effective way of producing the necessary skills and in particular should seek the optimal combination of general and specific training. In most cases, project identification precedes the analysis of a training specialist, who treats manpower and skill requirements as exogenously determined items and simply advises how to bridge the skill gap between existing and desired skills in the local labor force. As a result, little consideration may be given to whether alternative combinations of training and skills would be more efficient.

Where the general literacy base of a country is sufficient, according to Mingat, it may be profitable to provide more training than a simple calculation of skill requirements would justify, in view of the high rate of return to project-related training. In some projects, training for local workers may undoubtedly be more profitable (in terms of the final success of the project) than technical assistance, which usually accounts for a higher proportion of total project costs.

The importance of training was confirmed in another study (World Bank 1982) of investment projects in the agriculture and rural development, transport, industry and power, urban development, and population sectors. Not all the training needed for these projects could be provided within the formal education system. Other options were:

- The use of foreign experts, which is often the best means of speedily introducing new and fundamentally imported techniques, but involves high costs
- The use of local experts, who are less costly than foreign experts, may have better knowledge of local conditions, but may lack necessary technical expertise
- Fellowships for formal education or practical study abroad, which have the advantage of providing opportunities for firsthand observation of foreign techniques, but may expose the fellow to an environment unlike the one for which he is being trained
- The establishment of a formal training program locally to provide permanent training capacity, which may be needed if the manpower requirements of the project are substantial and recurring
- The strengthening of any existing training program, which may be possible if training is already well established in related subjects.

The review of project-related training concluded that the Bank had invested in a wide variety of types, amounts, and levels of training in order to meet individual project needs and to fulfill the Bank's commitment to institution-building and to developing the capacities of individuals and agencies on which development depends. Experience with these and follow-up projects in the countries concerned has given rise to substantial improvements in the Bank's work in training. The review (World Bank 1982, p. iv) concluded that during the 1970s

the Bank progressively broadened its project implementation training concerns to a wider development plane, made progress in preparing guidelines for training, and evinced more interest in and awareness of the role of the training function in human resource development. Nevertheless, although major lessons may have already been learned, the evidence also points to the need for further improvement. More systematic, concentrated and coordinated attention to training and human resource development is still needed if Bank operations are to make their intended contribution to member countries.

The important role of education and training in contributing to projects in other sectors is now recognized and being more fully exploited as part of the Bank's development policy. The 1980 *World Development Report* recommended "greater focus upon the human factor in national development," but stressed that this meant a change in emphasis rather than an increase in lending. The early projects of the World Bank tended to concentrate on the hardware aspects of projects, but during the 1970s more attention was paid to the identification of training needs, and a series of checklists were prepared, for example, for water supply and sewerage projects in 1976 and power and transportation projects in 1977.

These checklists for appraisal and supervision missions emphasized the need to check whether local counterparts were being trained to reduce dependence on foreign experts and also recommended systematic evaluation of training programs. Gradually, too, it was recognized that a broader approach to training was needed and that a project-by-project approach to manpower development was often too narrow. In 1980 a Bank statement on Policy and Guidelines on Training in Bank/IDA projects recommended:

- The context in which training requirements were assessed should not be based on narrow confines of project-related training, but should reflect wider sector considerations.
- The objectives of training should be expanded beyond project implementation to include efforts to improve planning, financing, and other management functions, and to create a continuing capability to identify manpower development needs beyond the project implementation period.

- Research into manpower needs should start at the earlier stages of project preparation.

In other words, Bank policy on training for projects in other sectors has gradually become broader and is now more concerned with long-term needs and institution-building rather than with immediate, narrow, and short-term requirements of individual projects. In addition, if the policy recommendations of the latest analysis of project-related training are accepted, Bank policy will become more concerned with comparing alternative training strategies in the design as well as the appraisal of projects.

Experience with Bank-financed projects underlines the need for concentrated preparatory work that will identify training needs, integrate the training program into the overall project so that training becomes a central focus rather than a peripheral addition, and incorporate evaluation procedures to ensure not only that the content and methods of training are adequate, but also that trainees do contribute to the success of the project.

Experience also suggests that greater use could be made of fellowships as a means of training in order to reduce reliance on foreign experts, and that the training component of many projects could be increased. More adequate support in the form of instructional materials for training programs would often increase their effectiveness.

A failing of projects has been the lack of dialogue between personnel in the education sector and in other sectors and insufficient emphasis on strengthening local education and training institutions, especially at the higher levels. In many cases, however, staff were trained for a particular project and in turn provided training at a later stage for other projects. In the Philippines, for example, the Water Management Technicians Training Center, set up as part of a specific project, has become a model for subsequent training programs in the Philippines. As a result of another project, the Liberia Electricity Corporation has strengthened its training machinery by rewriting the job descriptions of all senior staff to include an obligation to help in the training of all staff working under their direction. A good planning unit has been established as well, perhaps because of the initial impetus provided by a Bank-financed power project.

These examples show how the benefits of project-related training may be expanded and may gradually spill over into other sectors. They suggest, too, that the complementarities between training and the formal education sector are so important that all forms of human resource development, including general and specific training, must be given a high priority and central focus in the design and evaluation of projects in other sectors.

References

Anker, Richard. 1978. An Analysis of Fertility Differentials in Developing Countries. *Review of Economics and Statistics* 60(1).

Becker, Gary S. 1960. An Economic Analysis of Fertility. In National Bureau of Economic Research, *Demographic and Economic Change in Developed Countries*. Princeton, N.J.: Princeton University Press.

———. 1981. *A Treatise on the Family*. Cambridge, Mass.: Harvard University Press.

Bulatao, R. A. 1983. Mortality Thresholds for Fertility Transition in Developing Countries. World Bank, World Development Report Unit. Processed.

Caldwell, J. C. 1982. *Theory of Fertility Decline*. New York: Academic Press.

Cochrane, Susan H. 1979. *Fertility and Education: What Do We Really Know?* Baltimore, Md.: Johns Hopkins University Press.

Cochrane, Susan H., Donald O'Hara, and Joanne Leslie. 1980. *The Effects of Education on Health*. World Bank Staff Working Paper no. 405. Washington, D.C.

Colclough, Christopher. 1980. *Primary Schooling and Economic Development: A Review of the Evidence*. World Bank Staff Working Paper no. 399. Washington, D.C.

Feachem, R. G., David J. Bradley, Hemda Garelick, and D. Duncan Mara. 1983. *Sanitation and Disease: Health Aspects of Wastewater Management*. London: John Wiley.

Gwatkins, Davidson R. 1979. The End of an Era: Recent Evidence Indicates an Unexpected Early Slowing of Mortality Declines in Many Developing Countries. Washington, D.C.: Overseas Development Council.

Haveman, R., and B. Wolfe. 1984. Education and Economic Well-Being: The Role of Non-Market Effects. *Journal of Human Resources* 19(3):377–407.

Heyneman, Stephen P. 1983. Improving the Quality of Education in Developing Countries. *Finance and Development* 20, no. 1 (March):18–21.

Isenman, Paul, and others. 1982. *Poverty and Human Development*. New York: Oxford University Press.

Jallade, Jean-Pierre. 1982. Poverty Alleviation and Education Lending: Present Practice and Future Prospects. Washington, D.C.: World Bank, Education Department.

Jamison, Dean T., and Laurence J. Lau. 1982. *Farmer Education and Farm Efficiency*. Baltimore, Md.: Johns Hopkins University Press.

Mingat, Alain. 1984. Measuring the Economic Efficiency of Project-Related Training: Some Evidence from Agricultural Projects. Washington, D.C.: World Bank, Education Department.

Noor, Abdun. 1981. *Education and Basic Human Needs*. World Bank Staff Working Paper no. 450. Washington, D.C.

Schultz, T. 1964. *Transforming Traditional Agriculture*. New Haven, Conn.: Yale University Press.

Somerset, H. C. A. 1982. Examinations Reform: The Kenya Experience. Washington, D.C.: World Bank, Education Department.

Streeten, Paul, with Shahid Javed Burki, Mahbub ul Haq, Norman Hicks, and Frances Stewart. 1981. *First Things First: Meeting Basic Human Needs in the Developing Countries*. New York: Oxford University Press.

Tan, Jee Peng, and Michael Haines. 1984. Schooling and Demand for Children: Historical Perspectives. World Bank Staff Working Paper no. 697. Washington, D.C.

ul Haq, Mahbub, and Shahid Javed Burki. 1980. *Meeting Basic Needs: An Overview*. Washington, D.C.: World Bank.

Unesco. 1970. *Statistical Yearbook*. Paris.

———. 1977. *Statistical Yearbook*. Paris.

World Bank. 1980. *World Development Report 1980*. New York: Oxford University Press.

———. 1982. Review of Training in Bank-Financed Projects. Operations Evaluation Department Report no. 3834. Washington, D.C.

11

Conclusions

World Bank lending for education and training rose from $9 million in 1963 to nearly $1 billion in 1983. Although total spending represents less than 10 percent of the Bank's total investment in development, since the early 1960s investment in education and in project-related training has had a central role in the Bank's strategy of human development and the attack on poverty in the developing world. The Bank recognized in 1963 that spending on education is not simply consumption but is investment in human capital, and that education is not only a basic human right but also a means of enhancing the productive capacity of developing countries and increasing the profitability of investment in physical capital and basic infrastructure. This stance has been amply justified during the past twenty years, both by the practical experience of Bank-financed projects and by research.

Attempts to measure the contribution of education to economic growth (see chapter 2) suggest that a significant share of the growth of national income in developing countries is due to the education of the labor force, and that the contribution is much greater in Africa, where the stock of human capital is relatively small, than in Europe or Latin America. Not only do developing countries with the fastest rate of economic growth have higher rates of literacy than other countries at the same income level (Hicks 1980), but this is more likely to be due to the effect of education on income than vice versa, if we allow for the simultaneous effect of various factors (Wheeler 1980).

The results of such global comparisons are confirmed by microstudies that have examined the direct contribution of education to farmer productivity in Nepal and Thailand (Jamison and Lau, 1982), which show that education does indeed help to raise incomes and increase levels of production. Unlike the early research on the economic contribution of education, such evidence (see chapter 3) does not depend on wages and salaries as a measure of contributions to output, but is based on direct physical measurement of crop yields.

The research in this area has also demonstrated that education helps to increase the yield from investment in new agricultural techniques, improved irrigation, or the use of new seeds and chemical fertilizers. Farmers in Thailand, for example, are much more likely to use chemical inputs if they have had four years of primary schooling. The complementary nature of investment in education and investment in agriculture or rural development strengthens the case for investing in basic education as a way of raising the incomes of the poor. According to a recent analysis of fifty-two Bank-financed agricultural projects (Mingat 1984), a strong link exists between the size of the training component and the economic success of the project.

The External Advisory Panel on Education reporting to the World Bank in 1978 therefore concluded that the Bank has been correct in making substantial loans for the direct purpose of expanding and improving educational opportunities. The panel also agreed that education and training should not be regarded as a sector of development parallel with agriculture or industry, but should be considered essential elements of all organized efforts to speed up development.

Lessons from Experience

Thus the first lesson from the Bank's experience of investment in education and research on the contribution of education to development is that educational investment is productive, and that it does contribute directly to the growth and employment goals of developing countries. It also contributes indirectly by improving levels of health and life expectancy and by reducing fertility. There is now ample evidence that schooling fosters the type of behavioral change that is conducive to economic growth. It is also a vitally important complement to physical investment projects. Not only does investment in highways, power plants, irrigation, or industrial projects give rise to increased demand for both technical skills and general education, but such investments are more likely to yield high returns if accompanied or preceded by educational investment.

The second general lesson from two decades of experience of educational investment is that such investment must be broad rather than narrowly specialized, and that basic education, including both primary and nonformal education, is just as important for development as high-level technical education. The growing body of evidence in support of this conclusion has been summarized by the Bank's external advisory panel (World Bank 1978, p. 5):

General education is at least as important to development objectives as specific skill training and is, in fact, a necessary complement to the latter . . . the complementarity of general education and skill training is essentially related to the rapid changes being experienced by those who live in developing countries. In such a world a combination of general education and skill training is necessary to qualify persons to be able to adapt to change and to take part in it usefully.

The shift in Bank lending away from bricks and mortar and hardware to improvements in quality—for example, through curriculum development or the provision of textbooks—not only reflects the growing concern in developing countries about qualitative issues as well as quantitative expansion, but is also consonant with the evidence that the returns to qualitative improvements may be larger than the returns to quantitative expansion (Behrman and Birdsall 1983). That such a shift was necessary is borne out by the evidence that the quality of schools is vitally important. Contrary to the pessimism engendered by research in the United States, where the quality of schooling appears to be less important than family background in determining standards of achievement, the quality of school inputs in developing countries greatly influences the quality of output. Indeed, school factors are much more important than socioeconomic factors in determining levels of achievement and learning in developing countries, whereas family background appears to play a more important role in more advanced countries (Heyneman and Loxley 1983).

The answer to the question "Is it worth investing in education?" is therefore resoundingly positive. The question "Is it more profitable to invest in human skills and capacities or in physical capital?" is misguided, however, since it assumes that investment in people and investment in machines are alternatives. Investment choices should be concerned with the question of *how* rather than *whether* education can contribute to growth. Their proper concern is which particular combinations of general and technical education, which subject balance, which teaching methods, and which distribution of resources and opportunities for schooling would allow education to have the maximum impact on development and would best complement investment in physical capital and infrastructure. These are the questions that have been discussed in this book, but, as we have seen, there are no simple answers. Every investment proposal must therefore be evaluated with great care to determine its internal and external efficiency, its effect on the distribution of educational opportunities and the implications for equity, and its possible impact on other sectors.

Thus the remainder of this chapter does not recommend simple solutions or techniques to guide investment choices, but provides a brief summary of the main issues that must be taken into account before an educational investment is decided upon.

The External Efficiency of Educational Investment

The external efficiency of educational investment is usually judged by two criteria:

- The extent to which schools, universities, or training institutions provide the necessary skills for the smooth running of the economy, and the extent to which school-leavers or graduates are absorbed into the labor market, find the jobs and the earnings they expect, and are able to use their skills in employment
- The balance between the costs of investment in education and the economic benefits as measured by the higher productivity of educated workers, that is, by the social rate of return.

The early emphasis both in the World Bank and in developing countries on methods of forecasting skilled manpower requirements has given way to careful analysis of manpower patterns, including the way educated workers are used in the labor market. As a result, more emphasis is now placed on tracer studies, which follow the progress of the educated in the transition from school to work, than on methods of forecasting—which assume rigid relationships between manpower and output levels—or firm-based studies of the modern sector—which exclude the rural labor market, the self-employed, and the unemployed. Tracer studies of whole cohorts of graduates have been used to record the length of time taken to find a first job, and the success in finding a job has become one criterion of the success of an education project in meeting broad social and economic objectives. In a tracer study of secondary-school graduates in Indonesia (Clark 1983), for example, much of the educated unemployment has been shown to consist of an extended period of job search; virtually all secondary-school graduates eventually find employment and obtain higher earnings than those without secondary schooling. Consequently, the fact that secondary-school leavers may experience unemployment does not necessarily indicate an oversupply of secondary education.

Such an analysis of labor markets in developing countries should be included in the analysis of manpower patterns and trends, and the latter should be a continuous process rather than a forecasting exercise. This process could well include a cost-benefit analysis of the economic returns to education. Despite the well-known methodological objections to cost-benefit analysis (see chapter 3), investment appraisal should take place in the framework of a systematic comparison of the costs of different types or levels of education and the financial benefits measured in terms of productivity or earnings differentials. The World Bank's review of such analysis (Psacharopoulos 1981) leads to a number of general conclusions:

- The social returns to investment in education are still substantial, particularly in developing countries with a low stock of human capital.
- The returns to primary education are higher relative to highly specialized secondary or higher education, which has very high costs in developing countries but modest benefits with respect to lifetime earnings.
- Since the private returns are considerably higher than the social returns, education is still highly profitable for families and for individual students, and therefore the question is how educational investment should be financed.

Such conclusions do not, of course, provide an unambiguous guide for investment decisions. Rather, they identify the factors that should influence the choice of investments, and they emphasize the need for further research, for example, into the precise ways in which education increases the productivity of workers. World Bank research on the influence of education on farmer efficiency has pointed to the importance of general literacy and numeracy in determining the choice of agricultural techniques, for example, but detailed studies of the precise effects of education on productivity of workers in industrialized sectors and in self-employment are still lacking.

Since the social returns to education depend on the relation between costs and financial benefits, any policies that improve the internal efficiency of education and help to reduce costs, or that help to recover costs and shift some of the burden of financing educational investment from public to private sources, will increase the profitability of education as a social investment and make the distribution of benefits and the sharing of costs more equitable.

The Internal Efficiency of Education

Extensive research on the relation between school inputs and the level of pupil or student achievement (that is, the relation between input and output in educational institutions) has shown that student performance in developing countries is largely determined by the quality of school inputs and not by external socioeconomic factors. That means it is possible to improve internal efficiency by such measures as providing students with textbooks or improving teacher quality. Evidence from Chile, Peru, the Philippines, and Uganda (Heyneman, Farrell, and Sepulveda-Stuardo

1981; Heyneman, Jamison, and Montenegro 1984) points to the importance of school textbooks and shows that improving the availability of
textbooks may be one of the simplest and most cost-effective ways of
improving school efficiency.

The quality of teachers is another vital determinant of pupil performance (Husen, Saha, and Noonan 1978), and it is generally more important than class size (Haddad 1978). This finding raises the question of the
cost-effectiveness of alternative methods of teacher training. In-service
training and upgrading of teachers may often be a quicker and more
efficient way to improve teacher quality than enlarging initial training
capacity. Extensive analysis of distance-teaching methods (Perraton
1982) has demonstrated that radio, and in some cases television, may
increase the internal efficiency of education by producing lower rates of
wastage and repetition and lower unit costs. The costs of this new educational technology have often been underestimated, however. Furthermore, evaluations of new media may be subject to bias since the
advocates of new technologies tend to give them the benefit of the doubt.
This is not to suggest that new media do not have an important part to
play in improving the internal efficiency of education, but simply that all
alternative investment projects must be subjected to careful cost-
effectiveness analysis.

The Costs and Financing of Education

Overall, the rapid increase in educational expenditure that took place
in the 1960s and 1970s has slowed down in both developed and developing countries, whereas the pressure for expansion and qualitative improvements continues unabated. Thus an urgent need has arisen for a
reappraisal of both the patterns of costs and the financing of education.
Because the costs for a university student in an advanced country average
three to five times as much a year as for a primary-school pupil, whereas
in developing countries the ratio is 20:1, 40:1, 80:1, or even 100:1, poor
countries that invest heavily in higher education must shoulder an enormous burden. When unit costs of education are compared with national
income, the financial burden is even heavier. Despite the growing concern about this burden, there is no simple way to reduce costs. Only the
most careful scrutiny of costs in relation to what is achieved will ensure
that developing countries get the best value from scarce resources.

In addition to seeking ways to reduce the costs of education, many
countries are attempting to shift more of the costs from public to private
sources. Indeed, more attention should be given to the various methods

of cost recovery, including the use of fees for tuition or for meals and accommodation, student loans in place of scholarships or grants, and contributions from employers to help finance vocational education and training. These changes will not, by themselves, solve the financial constraints limiting educational investment, but they may help governments finance expansion or improvements that at present cannot be supported because of competing claims on public funds. Up to now, such shifts from public to private finance have by and large been resisted on the grounds of equity, but this opposition may not be justified (Mingat and Psacharopoulos 1985).

Equity Implications of Educational Investment

The equity implications of investment policy have attracted increasing attention since the early 1970s. As a result, it is now widely believed that educational investment must be evaluated on the basis of the following two criteria: the distribution of educational opportunities and facilities between different social groups, geographical areas, or rural and urban populations; and the distribution of financial burdens and benefits of education.

The rapid expansion of educational investment in developing countries in the 1960s and 1970s tended to benefit higher-income families and urban rather than rural communities and thus may have widened rather than reduced income disparities. The problem is that educational investment cannot, by itself, equalize incomes and employment opportunities. Nevertheless, it can help to raise the incomes of the poor if attention is paid to the social and geographical distribution of educational investment. Thus an effort must be made to increase the quality as well as the quantity of educational provision for the poor, for rural communities, for females, and for all who at present are underrepresented in education.

Recent World Bank analysis of the equity implications of alternative investment policies suggests that fees for tuition or for food and accommodation may, contrary to traditional beliefs, actually improve the equity of educational finance if the public funds saved by this means are then used to increase selective subsidies for the poor or to increase educational provision or the quality of schooling for disadvantaged groups. The introduction of student loans, which would reduce the extremely high level of subsidy for university education in many developing countries, could also have a positive distributional impact, since at present the high-income students are the ones most likely to benefit from education subsidies. The introduction of student loans, which a recent study

(Woodhall 1983) has shown to be feasible in developing countries, could therefore free public funds for greater expansion of primary education, which may well achieve both equity and efficiency objectives.

Although some policies may contribute to both equity and efficiency objectives, in some cases these goals may conflict, and it may be necessary to consider tradeoffs between efficiency and equity.

Links between Investment in Education and Other Sectors

The World Bank has provided considerable support for the training components associated with investments in physical infrastructure and in sectors other than education (such as transport, agriculture, mining, fuel and power). In the past, such investments in project-related training have been treated as relatively minor aspects of investment projects, but experience has shown that the training component not only plays an important part in determining the profitability of investment in physical capital, but it may also affect the general educational infrastructure and produce benefits that spill over into other sectors. In other words, investment in physical capital and infrastructure will not achieve its full potential without investment in the people who are ultimately responsible for the successful operation of that physical capital (Mingat 1984).

There is also growing evidence of the importance of links between educational investment and other types of investment in human capital, notably improvements in health and nutrition and reductions in fertility. Not only does educational investment help to reduce child malnutrition and mortality, but it contributes to general improvements in life expectancy and in the long run helps to reduce fertility. Because the relation between education and fertility is complex, however, an increase in educational investment will not automatically lead to a decline in the birth rate. Recent analysis shows, for example, that female education has more influence on fertility than male education (Cochrane 1979). Thus, measurement of the benefits of educational investment should take into account not only the direct benefits of higher labor market earnings of educated workers, but also the indirect effects on women's nonlabor market activities. Particularly important in this regard is the contribution of female education to improvements in the preschool abilities of children, which in turn may contribute significantly to educational efficiency (Selowsky 1976, 1982).

The Balance of Educational Investment

The general conclusion from two decades of World Bank experience of educational investment is that education fulfills a number of vital objectives:

- It satisfies a basic human need for knowledge, provides a means of helping to meet other basic needs, and helps sustain and accelerate overall development.
- It provides essential skilled manpower for both the industrialized and informal sectors of the economy, provides the means of developing the knowledge, skills, and productive capacities of the labor force, and acts as a catalyst in encouraging modern attitudes and aspirations.
- It helps to determine not only the incomes of the present generation but also the future distribution of income and employment.
- It influences social welfare through its indirect effects on health, fertility, and life expectancy and helps to increase the profitability of other forms of social and physical investment.

Because of the wide variety of benefits of educational investment and the diverse needs of different developing countries, no single formula can be applied to all countries or to all education projects. The gradual shifts that have occurred in World Bank lending for educational investment (see chapter 2) reflect the widespread belief that universal rules cannot be applied and that educational investment must be broad and diverse if it is to respond to the diverse needs of developing countries. In other words, the strategy of choice between alternative investments must be country-specific, although it must follow broad criteria.

The shift in allocations by level of education reflects the realization that it is not only secondary and high-level technical education that contributes to economic development, but that primary and nonformal education and training are also productive investments. The shift from investment in construction to the provision of equipment, learning materials, curriculum development, and technical assistance indicates that the qualitative aspects of education are now considered to be as important as quantitative expansion. Furthermore, the increase in lending for planning and managment reflects the Bank's resolve to develop managerial capacity and to involve developing countries more directly in determining educational priorities and evaluating alternative projects. The further achievement of that goal is the principal objective of this book.

References

Behrman, Jere R., and Nancy Birdsall. 1983. The Quality of Schooling: The Standard Focus on Quantity Alone Is Misleading. *American Economic Review* (December).

Clark, David H. 1983. *How Secondary School Graduates Perform in the Labor Market: A Study of Indonesia.* World Bank Staff Working Paper no. 615. Washington, D.C.

Cochrane, Susan H. 1979. *Fertility and Education: What Do We Really Know?* Baltimore, Md.: Johns Hopkins University Press.

Haddad, Wadi D. 1978. *Educational Effects of Class Size.* World Bank Staff Working Paper no. 280. Washington, D.C.

Heyneman, S., and W. Loxley. 1983. The Distribution of Primary School Quality within High and Low Income Countries. *Comparative Education Review* (February):108–18.

Heyneman, S., J. P. Farrell, and M. A. Sepulveda-Stuardo. 1981. Textbooks and Achievement in Developing Countries: What We Know. *Journal of Curriculum Studies* 13(3):227–46.

Heyneman, Stephen, Dean T. Jamison, and Xenia Montenegro. 1984. Textbooks in the Philippines—Evaluation of the Development Impact of Nationwide Investment. *Educational Evaluation and Policy Analysis* 6(2):139–50.

Hicks, Norman. 1980. *Economic Growth and Human Resources.* World Bank Staff Working Paper no. 408. Washington, D.C.

Husen, Torsten, Lawrence J. Saha, and Richard Noonan. 1978. *Teacher Training and Student Achievement in Less Developed Countries.* World Bank Staff Working Paper no. 310. Washington, D.C.

Jamison, Dean T., and Lawrence J. Lau. 1982. *Farmer Education and Farm Efficiency.* Baltimore, Md.: Johns Hopkins University Press.

Mingat, Alain. 1984. Measuring the Economic Efficiency of Project-Related Training: Some Evidence from Agricultural Projects. Washington, D.C.: World Bank, Education Department.

Mingat, Alain, and George Psacharopoulos. 1985. Financing Education in Sub-Saharan Africa. *Finance and Development* 22, no. 1 (March):35–38.

Perraton, Hilary. 1982. *Alternative Routes to Formal Education.* Baltimore, Md.: Johns Hopkins University Press.

Psacharopoulos, George. 1981. Returns to Education: An Updated International Comparison. *Comparative Education* 17(3):321–41.

Selowsky, M. 1976. A Note on Pre-School Age Investments in Human Capital in Developing Countries. *Economic Development and Cultural Change* 24, no. 4 (July):707–20.

———. 1982. The Economic Effects of Investment in Children: A Survey of the Quantitative Evidence. In *Child Development Information and the Formation of Public Policy: An International Perspective*, ed. T. E. Johnson. Springfield, Ill.: Charles C. Thomas.

Wheeler, David. 1980. *Human Resource Development and Economic Growth in Developing Countries: A Simultaneous Model*. World Bank Staff Working Paper no. 407. Washington, D.C.

Woodhall, Maureen. 1983. *Student Loans as a Means of Financing Higher Education: Lessons from International Experience*. World Bank Staff Working Paper no. 599. Washington, D.C.

World Bank. 1978. Report of the External Advisory Panel on Education to the World Bank. Washington, D.C.: Education Department.

Author Index

325

Subject Index

GEORGE PSACHAROPOULOS is education research adviser and manager in the Research Program, Education and Training Department, The World Bank.

MAUREEN WOODHALL is lecturer in educational administration, University of London Institute of Education, and was a consultant for the World Bank when this book was written.

The most recent World Bank publications are described in the annual spring and fall lists. The latest edition is available free of charge from the Publications Sales Unit, Department B, The World Bank, Washington, D.C. 20433, U.S.A.